W0111600

SpringerBriefs in Applied Sciences and Technology

SpringerBriefs present concise summaries of cutting-edge research and practical applications across a wide spectrum of fields. Featuring compact volumes of 50 to 125 pages, the series covers a range of content from professional to academic.

Typical publications can be:

- A timely report of state-of-the art methods
- An introduction to or a manual for the application of mathematical or computer techniques
- A bridge between new research results, as published in journal articles
- A snapshot of a hot or emerging topic
- An in-depth case study
- A presentation of core concepts that students must understand in order to make independent contributions

SpringerBriefs are characterized by fast, global electronic dissemination, standard publishing contracts, standardized manuscript preparation and formatting guidelines, and expedited production schedules.

On the one hand, **SpringerBriefs in Applied Sciences and Technology** are devoted to the publication of fundamentals and applications within the different classical engineering disciplines as well as in interdisciplinary fields that recently emerged between these areas. On the other hand, as the boundary separating fundamental research and applied technology is more and more dissolving, this series is particularly open to trans-disciplinary topics between fundamental science and engineering.

Indexed by EI-Compendex, SCOPUS and Springerlink.

Muhammad Faris bin Abd Manap ·
Solehuddin Shuib · Ahmad Zafir Romli

Total Hip Replacement (THR)

Effect of Inclination and Anteversion Angle on the Performance of Epoxy-UHMWPE Acetabular Cup

 Springer

Muhammad Faris bin Abd Manap
Mechanical Engineering Studies, Universiti
Teknologi MARA
Permatang Pauh, Pulau Pinang, Malaysia

Solehuddin Shuib
School of Mechanical Engineering
Universiti Teknologi MARA
Shah Alam, Selangor, Malaysia

Ahmad Zafir Romli
Faculty of Applied Sciences
Universiti Teknologi MARA
Shah Alam, Selangor, Malaysia

ISSN 2191-530X   ISSN 2191-5318 (electronic)
SpringerBriefs in Applied Sciences and Technology
ISBN 978-981-96-0974-1   ISBN 978-981-96-0975-8 (eBook)
https://doi.org/10.1007/978-981-96-0975-8

This Springer imprint is published by the registered company Springer Nature Singapore Pte Ltd.
The registered company address is: 152 Beach Road, #21-01/04 Gateway East, Singapore 189721,
Singapore

If disposing of this product, please recycle the paper.

Preface

The motivation behind this book stems from the critical challenges faced in Total Hip Replacement (THR) procedures, particularly concerning the performance and longevity of acetabular components. Despite advancements in implant technology, issues such as improper orientation and material failure continue to result in complications, including dislocation and wear. Recognizing the need for more reliable solutions, this book aims to explore the optimization of safe zone orientation and the development of composite materials, specifically focusing on Metal-on-Polymer (MoP) implants. By investigating these factors through numerical analysis, Finite Element Analysis (FEA) and experimental studies, the book seeks to enhance implant performance, reduce failure rates and ultimately improve patient outcomes in THR surgeries.

In 2005, I was invited by Prof. Amran bin Ahmed Shokri, a distinguished orthopedic surgeon from USM Kelantan, to attend a hip replacement surgery. This experience was profoundly enlightening, as I observed firsthand the meticulous precision required in the procedure. The surgery highlighted the complexity and challenges associated with hip implants, particularly in the orientation and material of the acetabular components. This observation sparked my motivation to explore this area further, especially given the limited research focused on optimizing the design and materials of hip and acetabular cup implants. This book is a culmination of that motivation, aiming to address the gaps in current research and contribute to improving surgical outcomes for patients undergoing total hip replacement.

The topic of hip implant design and its biomechanical performance is crucial due to the significant impact that total hip replacement (THR) surgery has on patients' quality of life. Hip implants are widely used to restore mobility and reduce pain in patients suffering from hip joint disorders. However, despite the advancements in THR, issues such as dislocation and failure of the acetabular components remain prevalent and can lead to severe complications, additional surgeries and increased healthcare costs.

This book is important because it addresses the underlying causes of these failures, particularly focusing on the dislocation issues that are among the top reasons for THR failure. By investigating the impact of stress at the contact region between the femoral head and the acetabular cup, as well as exploring suitable materials for the acetabular cup, the research aims to provide practical solutions that could significantly enhance the performance and longevity of hip implants.

The goal of this work is to contribute to the field of orthopedic implant design by offering insights that can lead to more reliable and durable hip implants. By improving the orientation of the acetabular cup and selecting appropriate materials, the study aims to reduce the likelihood of implant failure, thus improving patient outcomes. This research holds the potential to influence future designs of hip implants, offering better solutions to common problems and paving the way for more effective treatments in the field of orthopedic surgery.

We would like to extend our heartfelt gratitude to Prof. Dr. Amran bin Ahmed Shokri from the School of Health Campus, USM, for lending the hip implant model, which provided a clearer perspective on the THR analysis, and for his valuable advice on the current trends of THR in Malaysia. Our sincere thanks also go to Prof. Dr. Hazizan bin Md. Akil from the School of Material Engineering, USM, for generously providing us with samples of UHMWPE powders essential to our research.

During the experimental stage, we deeply appreciate the assistance provided by Mr. Muhamad Faizal bin Abd Halim, the laboratory research assistant, who played a crucial role in helping with the TrapeziumX program during the analysis of compression test results. We would also like to thank the Institute of Science (IOS) for their state-of-the-art equipment and laboratory facilities that were instrumental in our work.

We are profoundly grateful to the Research Management Centre (RMC), UiTM, and the Ministry of Higher Education Malaysia for their generous funding and unwavering support. This research would not have been possible without their invaluable contributions.

Our appreciation also goes to all the staff who provided the facilities and assistance during the testing and manufacturing processes. Their cooperation and willingness to assist were fundamental to the success of this study. A special thanks to our colleagues and friends who offered valuable insights, engaging discussions and helpful suggestions that have enriched this book.

Lastly, we sincerely thank our family for their unwavering support and encouragement throughout the preparation of this book.

This book is organized into nine chapters. Chapter 1 discusses the effect of inclination and anteversion angles on the performance of an Epoxy-UHMWPE acetabular cup. Chapter 2 reviews metal-on-metal total hip replacement and the safe zone orientation of the acetabular cup. Chapter 3 covers the finite element analysis (FEA) of metal-on-polymer (MoP) acetabular cup materials. Chapter 4 examines numerical and mathematical modeling for acetabular cup orientation. Chapter 5 explains the experimental work on Epoxy-UHMWPE materials. Chapter 6 presents the analytical results of safe zone orientation. Chapter 7 provides finite element analysis (FEA) static structural analysis. Chapter 8 details further experimental work on Epoxy-UHMWPE materials. Finally, Chap. 9 concludes by evaluating the performance of Epoxy-UHMWPE at various inclination and anteversion angles.

Conclusion

As I conclude this work, I reflect on the journey that brought me to this point—from observing meticulous surgeries to uncovering the nuances of hip implant design. This book is the result of years of dedication, collaboration and a relentless pursuit of knowledge. I am deeply grateful to my mentors, colleagues and family for their steadfast support throughout this endeavor. It is my sincere hope that the insights and findings presented in this book will contribute to advancing the field of total hip replacement, ultimately leading to better outcomes for patients around the world. Thank you to everyone who has been a part of this journey.

Thank you.

Shah Alam, Malaysia

Muhammad Faris bin Abd Manap
Prof. Ir. Ts. Dr. Solehuddin Shuib
Assoc. Prof. Dr. Ahmad Zafir Romli

About This Book

This book examines the critical factors influencing the success of total hip replacement (THR), focusing on safe zone orientation and the material composition of acetabular components. Through a combination of numerical analysis, finite element analysis (FEA) and experimental studies, the research identifies optimal safe zone orientations for Metal-on-Polymer (MoP) configurations, particularly highlighting the performance of different femoral head diameters. The study reveals that femoral heads larger than 28mm exhibit superior safe zone orientations, enhancing implant stability. Additionally, a new composite material, Epoxy-Ultra High Molecular Weight Polyethylene (EpUHMWPE5), is proposed for the acetabular cup. FEA results demonstrate that this material significantly reduces contact pressure, Von-Mises stress and total deformation, outperforming traditional UHMWPE at a 36mm femoral head diameter. This book provides valuable insights into optimizing THR procedures and materials, offering a path to improved implant performance and patient outcomes.

Contents

About the Authors

Muhammad Faris bin Abd Manap is a lecturer at Universiti Teknologi MARA (UiTM) Penang Branch, specializing in mechanical engineering. He holds a master's degree from Universiti Teknologi MARA (UiTM) and a bachelor's degree in mechanical engineering from Korea University. His research primarily focuses on the design and development of hip and knee implants, with an emphasis on finite element analysis (FEA), materials and biomechanical engineering. He is mainly recognized for his contributions to advancing implant technology and his commitment to academic excellence.

Ir. Ts. Dr. Solehuddin Shuib is a professor of biomechanics at the School of Mechanical Engineering, Universiti Teknologi MARA, Malaysia. He holds a B.S. in Mechanical Engineering from the University of Alabama at Birmingham, Alabama, USA, a master's in mechanical engineering from the University of Toledo, Ohio, USA and a Ph.D. in Mechanical Engineering from Universiti Putra Malaysia, Malaysia. He has authored or co-authored more than 70 refereed journal publications and presented at over 100 technical meetings. His research interests are in biomechanical engineering and medical devices.

Dr. Ahmad Zafir Romli is an associate professor at the Faculty of Applied Sciences, Universiti Teknologi MARA (UiTM), Malaysia. He graduated with a Ph.D. in Science from UiTM, specializing in the preparation and characterization of chicken feather/Epoxy composites. With a strong background in polymer science, Dr. Zafir has held several key positions at UiTM, including Program Coordinator for Polymer and Bio-composites, Head of the Center of Polymer Composites Research and Technology and Director of the Centre of Scientific Equipment. He has supervised numerous postgraduate students and published extensively in international journals and conference proceedings, contributing over 50 papers to the field. His research interests are centered on polymer composite systems, polymer testing and polymer processing, where he continues to advance the understanding and applications of polymer technology in various industries.

Abbreviations

ABD	Abduction
ADD	Adduction
ADL	Daily living activities
CCD	Neck-Stem Angle
CoC	Ceramic on Ceramic
ER	External Rotation
Erfl90	External Rotation at 90° flexion
EXT	Extension
FEA	Finite Element Analysis
FL	Flexion
HRA	Hip resurfacing replacement/arthroplasty
IR	Internal Rotation
IRfl90	Internal Rotation at 90° flexion
MoM	Metal-on-Metal
MoP	Metal-on-Polymer
OsA	Oscillation Angle
ROM	Range of Motion
SF	Synovial Fluid
THR	Total hip replacement

Symbols

α	Inclination angle
β	Anteversion angle
θ	Oscillation angle
A	Maximum angle of radius movement
n	Neck width at impingement level
r	Radius of head

a	True angle of femoral neck component position
b	Anteversion of femoral neck component around vertical axis
E	Foot of Perpendicular dropped from Start Point (B) to X-axis
D	Foot of Perpendicular dropped from Start Point (B) to Z-axis
c	Radial clearance
R_1	Radius of femoral head
R_2	Radius of acetabular cup
R'	Effective radius
E'	Effective elastic modulus
E	Young's Modulus
v	Poisson's ratio
F	Force exerted in Y-axis direction
δ	Displacement deformation
ρ	Density
m	Mass
V	Volume
μ	Coefficient of friction

List of Tables

Chapter 1
The Effect of Inclination and Anteversion Angle on the Performance of Epoxy-UHMWPE Acetabular Cup

1.1 Introduction

A normal native hip consists of a ball-and-socket joint that allows a wide range of motion to permit doing the activities of daily life (ADL). The ADL term is defined as the routine activities that people tend to do every day without needing any assistance. Transmission of forces can reach up to five times the body weight at the hip especially during running and climbing stairs activities and eight times during stumbling condition [1]. Muscle contraction, soft tissue tension and articular reaction forces will allow the balance of the external forces and moment exerted by the human body; thus, enabling the body to accommodate the high forces applied [2]. The femoral head usually forms two-thirds of a sphere with the other parts connected to the femur bone as shown in Fig. 1.1. Meanwhile, the acetabulum is formed by three parts of ilium, ischium and pubis that are called the Y-shaped triradiate cartilage [3]. The Y-shaped triradiate cartilage is illustrated in Fig. 1.2. The joint capsule has a strong configuration which gives stability to the human upon various gait motion. Naturally, hip joints are lubricated by synovial fluid which can be considered an electrolyte solution that contains proteins, lipids and hyaluronic acid that is stored in the synovial capsule. This fluid is important for the articulation between the femoral head and acetabulum as it will decelerate heat built-up on the bearing surface and avoid damaging the area [4].

In the condition of ADL being unable to perform naturally, a surgical procedure called Total Hip Replacement (THR) will be executed. Total hip replacement (THR) is a surgical procedure that removes an unhealthy hip and replaces it with a new implant to continue doing the ADL. Figure 1.3 explains the hip joint and the types of replacement used. There are two types of surgical approach on removing the unhealthy hip joint which are total hip replacement (THR) and Hip Resurfacing Replacement (HRA). Both are still valid in medical field, yet it is revealed that THR has good clinical results and long-term prosthesis survival compared to HRA due to a faster rate of bone healing upon postoperative surgery [6]. THR consists of an

© The Author(s), under exclusive license to Springer Nature Singapore Pte Ltd. 2024
M. F. b. A. Manap et al., *Total Hip Replacement (THR)*,
SpringerBriefs in Applied Sciences and Technology,
https://doi.org/10.1007/978-981-96-0975-8_1

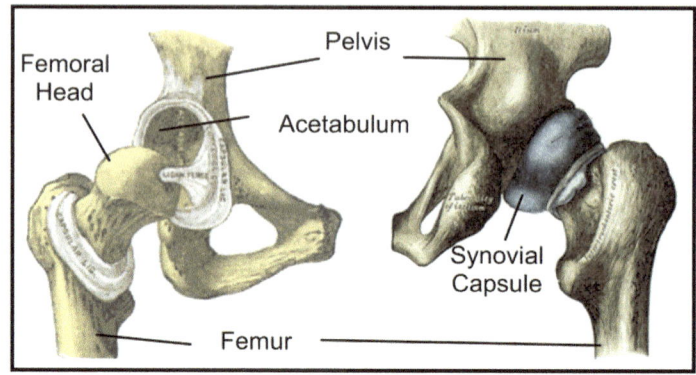

Fig. 1.1 Artificial Hip Implant Main Components (reproduced from [3], with permission)

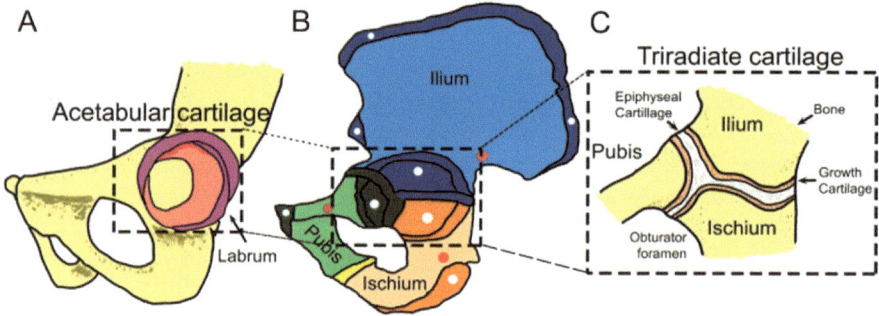

Fig. 1.2 The triradiate cartilage which is divided into ilium, ischium and pubis. **a** adult hemi pelvis, **b** A fetal hemi pelvis, **c** Triradiate cartilage (reproduced from [5], with permission)

average of three or four parts depending on the combination types that will be used as the implant. Conversely, HRA typically consists of only two main parts disregarding the femoral components parts. Metal, ceramics and polymers are common within material researches based on studies in hip implant. These materials are used individually or combined to get a desired acceptable range for any implant into the human body. Many techniques are still being developed today with the common combinations of metal with metal, ceramic with ceramic and metal with polymers. The notations Metal-on-Metal (MoM), Ceramic-on-Ceramic (CoC) and Metal-on-Polymer (MoP) are introduced by referring to the sequence of a femoral head and acetabular cup(liner), respectively [7].

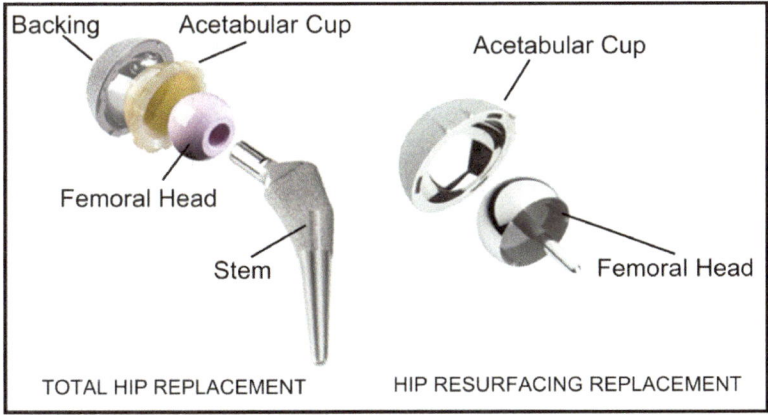

Fig. 1.3 The main components of total hip replacement and hip resurfacing replacement (reproduced from [3], with permission)

1.2 Problem Statement

The motivation for doing this research is based on the following research questions:

1. What possibly causes the dislocation issue among the top reasons for failure in THR?
2. How does stress affect the THR performance particularly at acetabular components contact region?
3. What kind of material should be considered at the acetabular cup that will avoid the failure of THR?

Hence, the problems upon THR will be discussed for a comprehensive understanding of this research. Hip prosthesis dislocation due to impingement is widely investigated and many researches are continuously conducted by worldwide researchers upon the issues [8–11]. Impingement is a mechanical abutment conflict between the bone of the femur and the pelvis. Specifically, in THR, impingement can be considered the unnecessary contact between the femoral head and acetabular cup that will cause poor outcomes after the surgery [12]. It is believed that lacking a degree in a range of motion of the hip prosthesis could lead to impinge and eventually cause dislocation [11]. Figure 1.4 illustrates the definition of impingement with primary and secondary effects. The unnecessary primary impingement contact happened at the edge of the acetabular cup where the femoral head contacts the rim of the acetabular cup. Conversely, the secondary impingement related to the femoral neck is in contact with the acetabular cup in an unordinary condition. A constant impingement will lead to the dislocation of the femoral head from the acetabular components.

A safe zone is introduced to find a suitable acetabular cup orientation that will be placed at the pelvis and commonly focuses on the inclination and the anteversion of

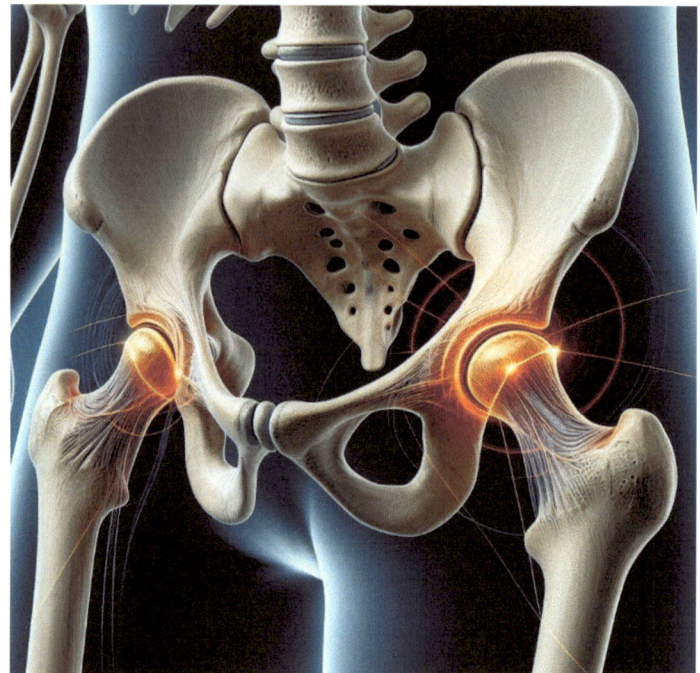

Fig. 1.4 Mechanism of impingement leading to dislocation (author's illustration)

the cup itself [9, 13, 14]. A case study showed that the so-called 'safe zone' orientation used since 1978 introduced by Lewinnek does not guarantee decreased risk of dislocation, yet the component orientation is still important upon doing THR [15]. A sufficient degree range of motion would give advantages to avoid impingement.

Besides the impingement-induced dislocation issue, the safe zone is also required to avoid edge-loading effect that proves to be clinically important and related to cup orientation [16]. Figure 1.5 shows the definition of the edge-loading that affects the superior cup region upon normal walking gait. At heel-strike of normal walking gait, with the hip flexed to approximately 30°, the load is concentrated approximately at the superior region [17]. There are also studies confirming that the liner acetabular cup had plastically deformed in the superior region, leading to hip instability [18].

The edge-loading effect is due to excessive contact pressure inside the articulate surfaces between the cup and femoral head. Creating a safe zone area which is a word that is invented upon this THR has also been known as important to reducing the edge-loading effect. The safe zone is an area that acetabular cup and femoral head orientation fit together without any possibility of severe damage in the future. The placement of the hip implant does need a safe zone even though a universal safe zone is nearly impossible to be measured as the design of the hip prosthesis itself acts as a pivot in determining the safe zone [19]. Dislocation is the common postoperative complication upon total hip replacement that ranges from 1% to nearly

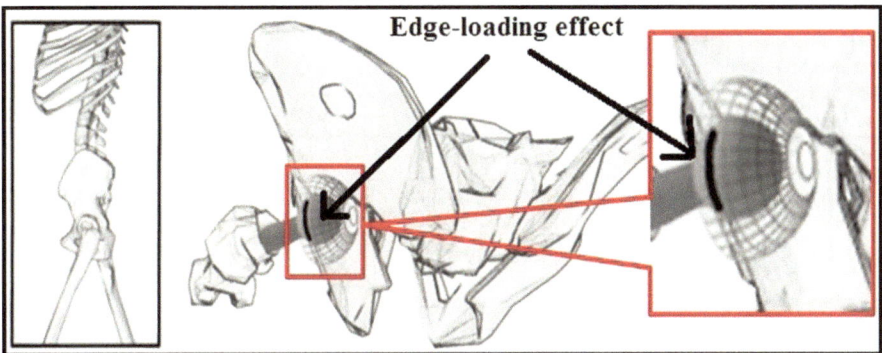

Fig. 1.5 Superior edge-loading during the normal walking gait is shown at the upper part of acetabular cup region (adapted and redrawn from [17], with permission)

10% in the United States and 17.9% in Australia [20, 21] with many factors leading to that situation. In that case, a new operation will be held necessarily to rectify the hip prosthesis for the patients and more complications look imminent.

Edge-loading causes one particular area to be exposed to concentrated stress; thus, causing deformation indentation and accelerating wear generated from the articulate surfaces. This wear will cause many issues particularly aseptic loosening and contribute to the implant failure, especially fracture [22]. It is not just the issue of the dislocation, as on impingement and edge-loading, but this orientation also gives complication to the wear effect, noise and cracking that is generated from the materials used in total hip replacement [23]. There are many types and combinations for the bearing joint and the entire decision-making on the suitability of the materials depends on the surgeon. These combination decisions are made based on their orientation and the patient's condition. In older patients, anteverted cup orientation is desired to achieve the degree of flexion compared to younger patients [24].

Acetabular cup made from UHMWPE is regarded as the best cup designed until today due to its biocompatibility and lower wear rate with a low coefficient of friction [25]. However, the wear rate is still considered higher than the previous use of PTFE cup and wear debris generated by UHMWPE triggers the osteoclast-mediated resorption and eventually, revision is needed [26]. At the time of writing, although UHMWPE was considered a biocompatible material, there is no fully biocompatible material available yet in the market. Due to the deformation with higher contact stress at the superior quadrant, the failure of the UHMWPE acetabular cup becomes rapid. Figure 1.6 shows that the load is coming into the bearing area corresponding to extensive stress in the rim of the cup; thus manifested in the form of cracks and the delamination of the rim interface [4]. The contact area of the acetabular cup with respect to femoral head needs to be defined appropriately to avoid excessive contact on one particular area only. The contact stress has been studied previously by using penetration depth concept that calculates the different cup inclination of UHMWPE and the results recorded that stress increased upon higher cup inclination angle with

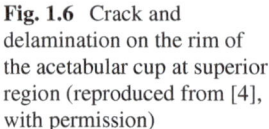

Fig. 1.6 Crack and delamination on the rim of the acetabular cup at superior region (reproduced from [4], with permission)

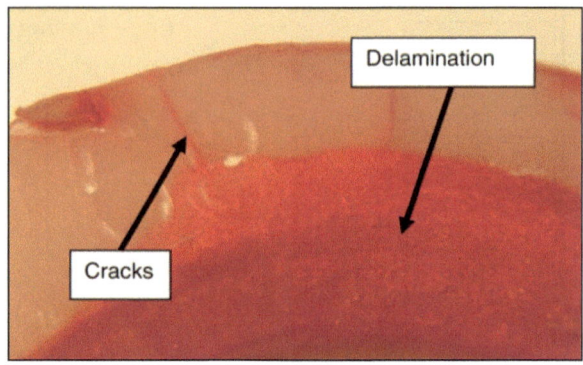

more than 65° radiographically [16]. Due to that, it is believed that tackling the issue on contact region area of acetabular cup with femoral head will improve the performance of the implant.

1.3   Research Objectives

1.   To find and propose a suitable range acetabular cup orientation to avoid dislocation and edge-loading effect.
2.   To study the mechanical properties of acetabular cup at different orientation angle and femoral head size by using finite element analysis (FEA).
3.   To suggest a new composite material that may enhance the mechanical performance of the articulate surface between acetabular cup and femoral head.

1.4   Scope of Research

The scope of research will be as follows:

1.   To find the safe zone orientation of acetabular cup by using a numerical modeling approach.
2.   To perform finite element analysis (FEA) by using the orientation taken from numerical modeling approach.
3.   To propose a new composite Epoxy/UHMWPE to replace the common acetabular cup material used.
4.   To obtain a new composite Young's Modulus value by using compression test experiment.
5.   To compare a new composite with the existing material used for the acetabular cup in simulation FEA.

1.5 Significance of the Study

Theoretically, by improving the orientation of the acetabular cup and materials properties used, it advances the performance of the implant to look more imminent which may avoid prosthesis dislocation and decrease the contact pressure [11, 27, 28]. The failure of the hip implant that occurs due to the contact region of the femoral head with acetabular cup could be reduced by this project. The orientation of the acetabular cup and the strength of the materials that will be used for the implant had been continuously studied in order to achieve a longer hip implant lifespan and better performance. It is hoped that by the end of this project, factors leading to acetabular component early failure due to the performance of the hip implant may be analyzed and a better solution could be attained in the future.

1.6 Book Outline

This book is organized into nine chapters. Chapter One discusses the effect of inclination and anteversion angles on the performance of an Epoxy-UHMWPE acetabular cup. Chapter Two reviews metal-on-metal total hip replacement and the safe zone orientation of the acetabular cup. Chapter Three covers the finite element analysis (FEA) of metal-on-polymer (MoP) acetabular cup materials. Chapter Four examines numerical and mathematical modeling for acetabular cup orientation. Chapter Five explains the experimental work on Epoxy-UHMWPE materials. Chapter Six presents the analytical results of safe zone orientation. Chapter Seven provides finite element analysis (FEA) static structural analysis. Chapter Eight details further experimental work on Epoxy-UHMWPE materials. Finally, Chapter Nine concludes by evaluating the performance of Epoxy-UHMWPE at various inclination and anteversion angles.

References

1. G. Bergmann, G. Deuretzbacher, M. Heller, F. Graichen, A. Rohlmann, J. Strauss, G.N. Duda, Hip contact forces and gait patterns from routine activities. J. Biomech. **34**(7), 859–871 (2001)
2. G.E. Lewinnek, J.L. Lewis, R. Tarr, C.L. Compere, J. Zimmerman, Dislocations after total hip replacement arthroplasties. J. Bone Joint Surg. Am. (60), 217–220 (1978)
3. L. Mattei, F. Di Puccio, B. Piccigallo, E. Ciulli, Lubrication and wear modelling of artificial hip joints: a review. Tribol. Int. **44**(5), 532–549 (2011)
4. N.D.L. Burger, P.L. de Vaal, J.P. Meyer, Failure analysis on retrieved ultra high molecular weight polyethylene (UHMWPE) acetabular cups. Eng. Fail. Anal. **14**(7), 1329–1345 (2007)
5. E. Oussoren et al., Hip disease in Mucopolysaccharidoses and Mucolipidoses: a review of mechanisms, interventions and future perspectives. Bone **143**, 115729 (2021)
6. S.K. Fokter, N. Fokter, T. Hospital, Hip fracture in the elderly: partial or total arthroplasty? in *Recent Advances in Hip and Knee Arthroplasty* (2012)
7. U. Holzwarth, G. Cotogno, *Total Hip Arthroplasty: State of the Art, Challenges and Prospects*, vol. 1 (2012)

8. K.-H. Widmer, B. Zurfluh, Compliant positioning of total hip components for optimal range of motion. J. Orthop. Res.: Off. Publ. Orthop. Res. Soc. **22**(4), 815–821 (2004)
9. F. Yoshimine, K. Ginbayashi, A mathematical formula to calculate the theoretical range of motion for total hip replacement. J. Biomech. **35**(7), 989–993 (2002)
10. M. Ghaffari, R. Nickmanesh, N. Tamannaee, F. Farahmand, The impingement-dislocation risk of total hip replacement: effects of cup orientation and patient maneuvers, in *Conference Proceedings : IEEE Engineering in Medicine and Biology Society*, vol. 34, no. 1 (2012), pp. 6801–6804
11. D. Kluess, H. Martin, W. Mittelmeier, K.-P. Schmitz, R. Bader, Influence of femoral head size on impingement, dislocation and stress distribution in total hip replacement. Med. Eng. Phys. **29**(4), 465–471 (2007)
12. A. Malik, A. Maheshwari, L.D. Dorr, Impingement with total hip replacement. J. Bone Jt. Surg. Am. Vol. **89**(8), 1832–1842 (2007)
13. K.-H. Widmer, A mathematical formula to calculate the theoretical range of motion for total hip arthroplasty. J. Biomech. **36**(4), 615 (2003)
14. C.L. Harrison, A.I. Thomson, S. Cutts, P.J. Rowe, P.E. Riches, Research synthesis of recommended acetabular cup orientations for total hip arthroplasty. J. Arthroplast. **29**(2), 377–382 (2014)
15. C.I. Esposito, B.P. Gladnick, Y. Lee, S. Lyman, T.M. Wright, D.J. Mayman, D.E. Padgett, Cup position alone does not predict risk of dislocation after hip arthroplasty. J. Arthroplast. **30**(1), 109–113 (2015)
16. X. Hua, B.M. Wroblewski, Z. Jin, L. Wang, The effect of cup inclination and wear on the contact mechanics and cement fixation for ultra high molecular weight polyethylene total hip replacements. Med. Eng. Phys. **34**(3), 318–325 (2012)
17. W.L. Walter, G.M. Insley, W.K. Walter, M.A. Tuke, Edge loading in third generation alumina ceramic-on-ceramic bearings: stripe wear. J. Arthroplast. **19**(4), 402–413 (2004)
18. D.A.J. Wilson, J.P. Corkum, M.G. Teeter, D.W. Holdsworth, M.J. Dunbar, Early failure of a polyethylene acetabular liner cemented into a metal cup. J. Arthroplast. **27**(5), 820.e5–820.e8 (2012)
19. K.-H. Widmer, Is there really a "safe zone" for the placement of total hip components? in *Bioceramics and Alternative Bearings in Joint Arthroplasty*, (2006), pp. 249–252.
20. H.-M. Ji, K.-C. Kim, Y.-K. Lee, Y.-C. Ha, K.-H. Koo, Dislocation after total hip arthroplasty: a randomized clinical trial of a posterior approach and a modified lateral approach. J. Arthroplast. **27**(3), 378–385 (2012)
21. AOA, Australian Orthopaedic Association National Joint Replacement Registry. in *Annual Report*, (2015)
22. X. Hua, J. Li, L. Wang, Z. Jin, R. Wilcox, J. Fisher, Contact mechanics of modular metal-on-polyethylene total hip replacement under adverse edge loading conditions. J. Biomech. **47**(13), 3303–3309 (2014)
23. A.B. Patel, R.R. Wagle, M.M. Usrey, M.T. Thompson, S.J. Incavo, P.C. Noble, Guidelines for implant placement to minimize impingement during activities of daily living after total hip arthroplasty. J. Arthroplast. **25**(8), 1275–1281.e1 (2010)
24. K.-H. Widmer and M. Majewski, The impact of the CCD-angle on range of motion and cup positioning in total hip arthroplasty. Clin. Biomech. (Bristol, Avon) **20**(7), 723–8 (2005)
25. S. Roy, S. Bag, S. Pal, In vitro biomechanical evaluation of UHMWPE and its composites as biomaterial. Trends Biomater. Artif. Organs **17**(2), 54–60 (2004)
26. G. Lewis, Properties of crosslinked ultra-high-molecular-weight polyethylene. Biomaterials **22**(4), 371–401 (2001)
27. S. Ge, S. Wang, X. Huang, Increasing the wear resistance of UHMWPE acetabular cups by adding natural biocompatible particles. Wear **267**(5–8), 770–776 (2009)
28. X. Hua, J. Li, Z. Jin, J. Fisher, The contact mechanics and occurrence of edge loading in modular metal-on-polyethylene total hip replacement during daily activities. Med. Eng. Phys. **38**(6), 518–525 (2016)

Chapter 2
Metal-On-Metal Total Hip Replacement and Safe Zone Orientation of Acetabular Cup

2.1 Introduction

This chapter will cover five main sections. The first section describes the trends in THR. The second section focuses on the safe zone orientation of acetabular cup. The third section mainly discusses the MoP THR using FEA. The fourth section covers the materials being used in acetabular cup. The final section will point out the summary of Chapter Two.

2.2 Total Hip Replacement

Typically, an artificial hip implant is designed for long-term use, but there are also cases where a revision is required. The revision was performed for the artificial implant that did not achieve its intended standards which required another surgery. There is a report claimed that revision occurred higher in patient under 55 years old in the region of the United Kingdom with the primary THR reported at 800,683 cases by the end of 2015 [1]. At the time of writing, the Australian Orthopaedic Association National Joint Replacement Registry THR data exhibits the main reasons for doing revision after THR. The summary of the data is shown in Fig. 2.1. This summary shows that osteolysis, loosening, dislocation and fracture are the highest rates of doing revision in Australia by 2015 [2]. Although many new techniques have been constantly developed throughout the year to minimize doing the revision, the THR failure still occurred. It is important that the main reasons of THR failure should be resolved, thus minimizing the revision surgery.

A prosthetic hip implant THR normally contains a long stem implanted in the femur and a spherical femoral head that articulates with the acetabulum [3]. The top of the thighbone (femur) is the largest bone joint with the horizontal pelvic coxal bone and lower end fixed at the knee [4, 5]. The area contains diarthrosis, also called

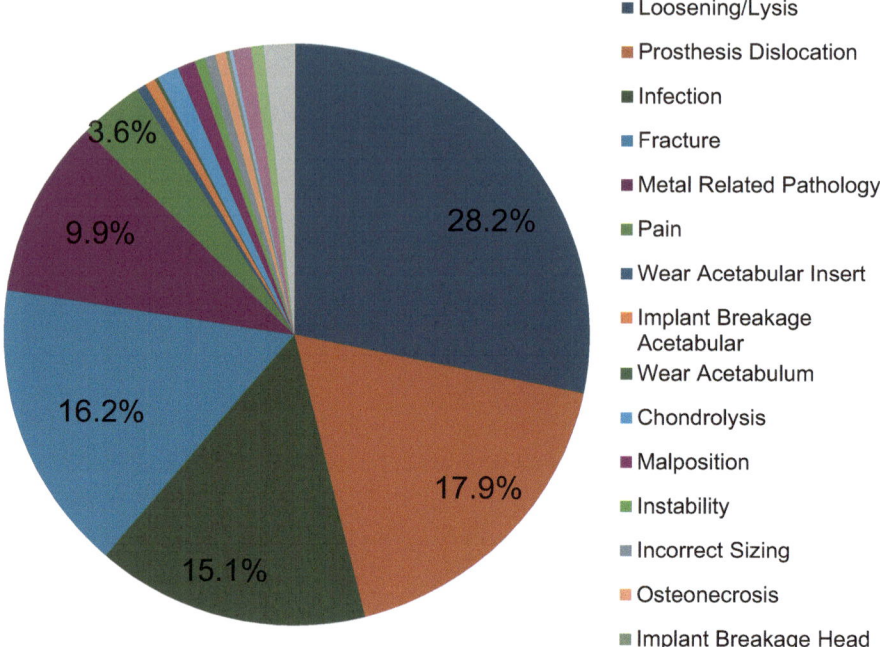

■ Loosening/Lysis
■ Prosthesis Dislocation
■ Infection
■ Fracture
■ Metal Related Pathology
■ Pain
■ Wear Acetabular Insert
■ Implant Breakage Acetabular
■ Wear Acetabulum
■ Chondrolysis
■ Malposition
■ Instability
■ Incorrect Sizing
■ Osteonecrosis
■ Implant Breakage Head

Fig. 2.1 The main reasons of THR failure recorded in Australia by 2015 (redraw from [2], with permission)

a synovial joint, since it is wrapped in a capsule that contains the synovial fluid (SF), a biological fluid that acts as a shock absorber [5]. The primary goal of THR is to mimic the six basic movements of the hip joint: flexion/extension, adduction/abduction and external/internal rotation [6]. A THR is considered successful if all six basic ranges of motion can be done by the patients following the surgery.

2.2.1   Metal-On-Polymer Total Hip Replacement

In MoP THR, there are two ways of doing THR, which are cemented and cementless hip implants. The definition of cemented hip implant means the femoral stem was dipped into the femur bone with bone cement used as the grouting material in between those materials [7] and the cementless is by coating the femoral stem with porous materials to encourage bone ingrowth at the implant [4].

Also at the bearing joint, cemented is defined by applying a single acetabular cup with a bone cement acting as a grouting material to the pelvis and cementless is defined by adding a metal backing/shell to the acetabular cup component to absorb dynamic loads, respectively [5]. The acetabular cup is defined with many terms, which are sometimes referred to as a cup, insert or liner, depending on individual preferences. Both types are still widely used in THR, as each combination has its own functionality and suitability for the patients. However, data shows that upon doing THR, it is more intended to use a metal backing (dual mobility) instead of using bone cement [8].

2.2.1.1 Ultra High Molecular Weight Polyethylene (UHMWPE) Acetabular Cup

In MoP THR, Ultra High Molecular Weight Polyethylene (UHMWPE) is common for the acetabular cup component. The introduction of this new polymer has revolutionized the THR industry as this material is considered a biocompatible polymer and the history of this material recognition was explained by Kurtz et al. [9] in their handbook. Before this, Polytetrafluoroethylene (PTFE) is well-known as the best material selection for doing MoP THR. Even though PTFE has been used earlier than the introduction of UHMWPE, the clinical assessment shows that UHMWPE gives better performance upon doing THR especially after they discovered that PTFE failed in vivo due to significant wear [9]. However, there is certain consideration that must be alerted such as the mechanical properties of PTFE are undeniably as good as UHMWPE, particularly the lower coefficient of friction on dry sliding that will smooth the articulate motion between the femoral head and acetabular cup [10]. Recently, Cross-Linked Ultra High Molecular Weight Polyethylene (XLPE) has been introduced, which was considered a modification of UHMWPE. The statistics show that XLPE has a significant effect on wear resistance although it loses some mechanical strength and becomes more brittle than UHMWPE [11]. However, it is also revealed that the results of XLPE are only based on the short-term studies without knowing any results on the long-term survival of the materials in THR [12].

Thus, the articulation of the acetabular cup and femoral head on this bearing system has been studied by prioritizing the leading factor of implant failure based on these materials at the acetabular region. Many materials have been studied in order to replace the UHMWPE acetabular cup with the aim of finding a new material replacement that has mechanical properties as good as UHMWPE. Epoxy, polyurethane (PU) and polytetrafluoroethylene (PTFE) are well-known for their usage as blood-contacting biomaterials and are considered biocompatible with the human body [13]. Table 2.1 shows the comparison of mechanical properties of UHMWPE, PTFE and PU that have been used previously for the acetabular cup component. The range of mechanical properties in terms of Young's Modulus and Poisson's ratio does not exhibit much difference among those three materials. It is supposed that introducing more of these kinds of materials might improve the acetabular components' failure, especially on the articulate surface motion on the bearing joints.

Table 2.1 The comparison of mechanical properties of UHMWPE, PTFE and PU that have been used in acetabular components

Acetabular Cup Materials	Young's Modulus (MPa)	Poisson's Ratio
UHMWPE [14, 15]	200 ~ 1300	0.35 ~ 0.47
PTFE [16, 17]	500 ~ 1400	0.46 ~ 0.49
PU [18, 19]	200 ~ 2000	0.28 ~ 0.50

2.2.1.2 Cementless and Bigger Femoral Head in MoP THR

Among the critical issues related to hip joints are chronic pain and diseases such as osteoarthritis, rheumatoid arthritis, bone tumors or traumas [5]. In these cases, the best solution is performing total hip replacement (THR), which is a surgical procedure of removing an unhealthy hip joint with an implant. The surgical procedure requires the head of femur to be removed totally and the acetabulum to be polished. The new acetabular cup component will replace the infected head that is being removed. For surgical procedures that use individual polyethylene, bone cement, typically Poly Methyl Methacrylate (PMMA), is used as an adhesive material between acetabular cup and acetabulum that has been polished.

The demerits of the implant with these material components have been discovered, mostly related to wear and aseptic loosening of the implant [20, 21]. Dual mobility, which implies metal backing instead of bone cement, has been utilized with the aim of increasing the stability of the implant itself. However, issues regarding initial fixation still arise with this combination [8]. For the time being, metal backing is considered the best solution for the case of using MoP combination in THR [8].

Meanwhile, at the femoral component, femoral stem is dipped into medullary canal of femur bone and commonly bone cement Poly Methyl Methacrylate (PMMA) is used as grouting material [7]. A preliminary report said that MoP hip implant is done on more than 100,000 European patients [22]. The amounts are expected to increase further due to an aging population, decreasing average age at the first operation and the limited life span of prostheses [11]. The rate of hip replacement increased by approximately a quarter percent between 2000 and 2009, thus this phenomenon is expected to continue in the next decade parallel with technological improvement in developing countries even though the procedure is considered expensive interventions [23].

There are four main components at MoP THR, which are the femoral stem, femoral head, acetabular cup and the fixation agent cement at the acetabulum and femoral part. However, considering the trends in this millennium year, there is much consideration of using the cementless components at both the femoral component and acetabular component [24, 25]. This statement means that the fixation agent cement for the acetabular component is replaced by the backing (usually metal) that will allow the ingrowth of the bones to the backing. Yet, the usage of cement as the fixation agent remains widely used because cementless fixation seems to have better growth only in younger patients compared to older patients [26, 27]. Cementless total hip arthroplasty is varied either by 'press-fit' technique or using screw technique. 'Press-fit' is meant by eliminating the screw usage as the fixation agent to the pelvis by allowing bone ingrowth to the porous surface of metal backing from the acetabulum. The trend of using cementless for both acetabular components and femoral components is at a higher rate compared to cemented components. There are some supported reports which mentioned that acetabular loosening is caused by the usage of cemented femoral component [27, 28]. It is suggested that the performance of fully cementless implant may be improving over time compared to cemented implants [29]. However, there are different opinions suggested in which a review paper shows that a fully cementless implant will increase the probability rate of doing revision after THR [28]. Later on, many surgeons applied the hybrid total hip replacement techniques, which means only acetabular components are cementless. This is because hybrid total hip replacement shows excellent results in 10 years of follow-up and it is also cost-effective compared to other techniques [30, 31].

The current conditions of the surgical procedure have an issue related to dislocation of hip implant, wear and tear of the acetabular cup, loosening phenomenon of the fixation agent and fracture of the femoral stem. PMMA cement is classified as the preferred material for the hip replacement models since it was developed by Sir John Charnley and society demanded it for improving lifespan of the THR, which established research awareness on this matter [32]. Since 1970, a great achievement in orthopedics with PMMA has been noted as a successful cement in THR which gave the fact that PMMA does not chemically bond with either bone or implant [33]. However, due to issues arising from the use of PMMA, it is common that metal backing almost eliminated the usage of PMMA cement as the fixation agent at the acetabular component in the case of using the hard-on-soft bearing connection.

Typically, there are different sizes used for the parts of acetabular components depending on the expert expertise and the suitability of the patients. However, most of the studies showed that any combination, whether it is MoM, CoC or MoP, is encouraged to use a larger femoral head to prevent dislocation and increase the ROM [34]. By the time of the writing, currently, there is no consensus on what constitutes a "small-diameter" or "large-diameter" head. Nevertheless, there is a suggestion that the femoral head diameter should be between 22mm and 28mm for small, 28mm to 36mm for medium, and more than 36mm for larger [35]. Figure 2.2 shows some typical sizes available in the market for the femoral head sizes. For the record, larger than 36mm femoral head diameter size is available in market, which is typically used in MoM and it is proven that it will increase the range of motion [22]. Previously, there are studies in vitro that indicate that femoral head larger than 32mm will increase the flexion angle and consequently minimize the dislocation [34]. Another study also shows that the dislocation possibilities are higher at femoral head diameter of 22mm compared to 28mm [36]. Based on these researches, there is suggestion that a bigger femoral head diameter may improve the implant life, especially upon reducing dislocation issues and increasing the range of motion.

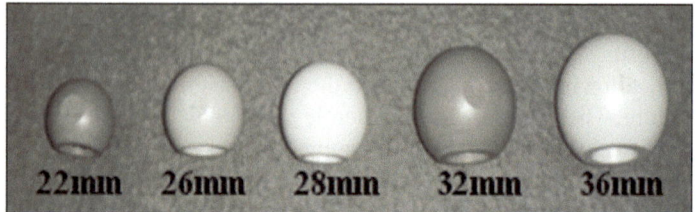

Fig. 2.2 Some common femoral head sizes available in the market (Reproduced from [37], with permission)

2.3 Safe Zone Orientation of Acetabular Cup

Previous researches had shown that there are some different orientations that define the suitable, so-called safe zone orientation for the placement of the acetabular cup. The aim of this orientation is to find a suitable range of motion that may avoid dislocation, impingement and edge-loading, resulting in excessive contact pressure at the bearing joint, especially at the superior cup area [38–42]. Most researchers intended to define the proper cup orientation based on the inclination angle and anteversion angle. These two main angles have the reference with respect to the human anatomy plane frame. Researchers have different reference frames upon finding the safe zone orientation of acetabular cup but these three reference frames are the common references in order to determine the orientation [43] as shown in Fig. 2.3. Meanwhile, Fig. 2.4 shows the schematic diagram of left hip joint with acetabular components; thus, the definition of inclination angle and anteversion angle that make out the so-called safe zone in terms of orthopedic field could be displayed.

Safe zone orientation is particularly an important criterion upon doing THR and the summarization of previous research on safe zone orientation recommended is shown in Table 2.2. In every case, researchers give a range of safe zones that are allowable at the acetabular component with own support judgment. They claim that if acetabular cup placement is within the safe zone orientation, the complications that may happen after the surgery can be avoided. The tabulated data shows a general assumption upon the acetabular orientation disregarding any method that was being

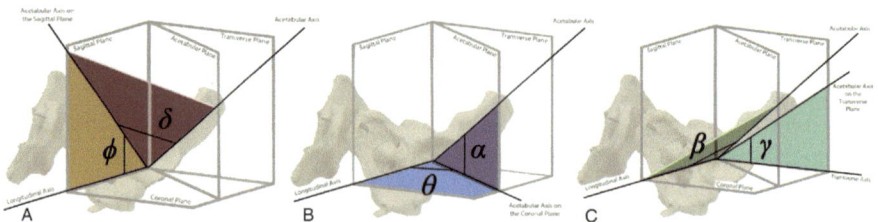

Fig. 2.3 Definition of the whole body and pelvic reference frame in the position upon surgery of THR. **a** Operative, **b** Radiographic and **c** Anatomical (reproduced from [44], with permission)

Fig. 2.4 Left hip joint shows the definition of inclination angle and anteversion angle represented by α and β, respectively (reproduced from [41], with permission)

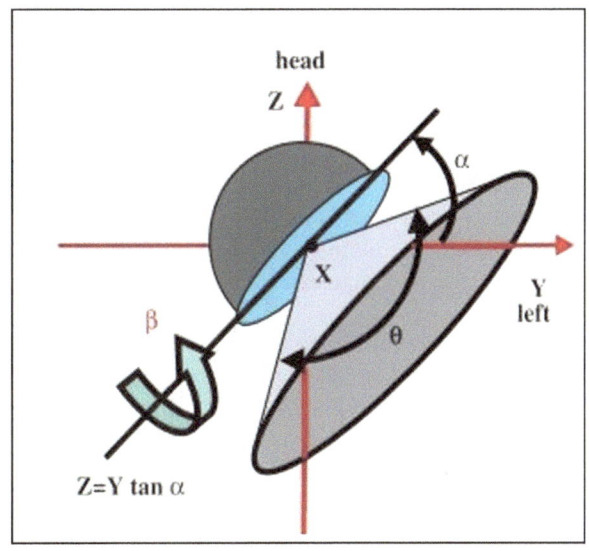

used either by theoretical or experimental analysis. In 2008, there was a research that intended to resolve inconsistency of the safe zone orientation of the acetabular cup and their assumption was that averaged recommendation is at 41° inclination angle matched with 16° anteversion angle disregarding specific parameters [43].

It is clearly stated that the safe zone orientation of acetabular cup could not be defined as a universal angle. The complexity and variation of recommendations that have been proposed show that many researchers have their own arguments on the

Table 2.2 Summary of recommended safe zone orientation based on previous research definition [43–45]

References	Inclination angle (°)	Anteversion angle (°)
Lewinnek et al. [46]	30–50	5–25
McCollum et al. [47]	30–50	20–40
Dorr et al. [48]	− 55	10–25(OA)
Seki et al. [49]	30–50	10–30
Jolles et al. [50]	50	20 (OA)
Widmer et al. [51]	40–45	20–28
Yoshimine et al. [52]	35–55	10–30
Biedermann et al. [53]	35–55	5–25
Kummer et al. [54]	35–50	0–10
Ko, Byung-Hoon et al. [44]	39–41	15–17
Mclawhorn et al. [45]	46.9–37.3	25

OA: An operative anteversion (the angle is defined as per operation condition).

best safe zone orientation. The data from previous studies suggested that researchers tried to make one general assumption range of safe zone orientation. Due to that, the acetabular cup orientation needs to be understood explicitly in order to get an overall view of how to define the orientation angle. Dislocation, edge-loading, impingement of hip joint and wear of acetabular components are among the desirable considerations for suitability of acetabular components in THR [41, 55]. The demand of doing the common activities of daily living (ADL) becomes pivot in avoiding impingement with ROM as the main criterion [38]. Turning the torso in the supine position and sitting on the low chair are known to be the frequent reasons of dislocation of hip implant [47, 49].

The common postures during ADL require combination of flexion/extension, adduction/abduction and internal/external rotation [6, 44, 56] as shown in Fig. 2.5. Acetabular orientation needs to be measured properly as it is particularly aiming to avoid edge-loading and prosthetic impingement in THR. Basic science studies [57, 58] state that it is recommended to position the acetabular cup radiographically at an inclination of < 50° in order to avoid edge-loading effect, which occurs when the femoral head makes contact with the acetabular component near the rim in loading conditions. Lewinnek et al. [46] discovered in the late 1970s that anterior dislocation occurred when acetabular component anteversion increased; thus they proposed a new range of safe zones that may reduce the dislocation probability.

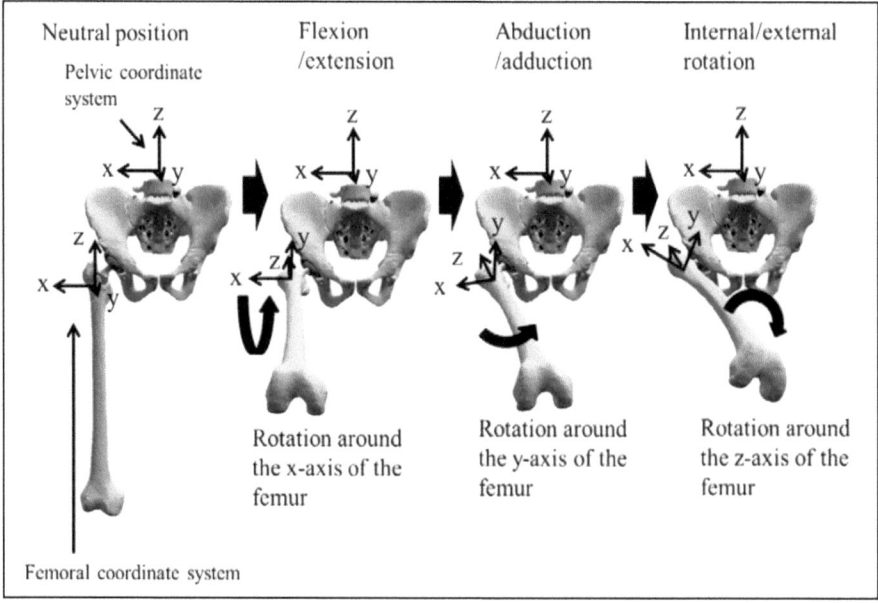

Fig. 2.5 The definition of hip Range of Motion (ROM) that is required after the THR surgery (reproduced from [59], with permission). The ROM is an important criterion for the calculation of a new safe zone orientation angle

Other researchers have also shown that increasing the range of motion (ROM) may permit common ADL and this fact is proven by the development of mathematical formulations to calculate optimum ROM for THR [56]. Strong indication stated that ROM played a significant role in minimizing the unintended poor effect that may consequence from THR.

Creating the safe zone to avoid prosthetic impingement was introduced by previous research [38, 46, 56] and it is defined as the area that can fulfill all the criteria needed for ROM to avoid impingement that causes dislocation. From the research recommendation, there is no clear indication that a universal orientation for cup safe zone can be concluded; thus the research is still open for new findings. However, there are some criteria that state the compliant positioning of acetabular cup needs four orientation parameters, i.e., cup inclination, cup antetorsion, stem antetorsion and stem CCD-angle. It was said if one parameter is fixed, then other parameters must follow the accordance set that is already established [60]. To show the importance of safe zone orientation, a 3D parameterized and visual kinematic simulation tool has been created to investigate a proper orientation upon a theoretical range of motion [61]. There are also clinical studies that compare real patient cases with theoretical safe zone orientation, and the results are disappointing because only a few implants are within the safe zone orientation [62]. It is a clear indication that safe zone orientation of the acetabular component is a top priority when considering THR. With the current trends, the research on the acetabular component orientation is still widely investigated not just depending on the current parameters. Most of them are trying to add more parametric studies when dealing with any proper safe zone orientation in order to get a more accurate safe zone orientation angle. Among them, researchers are considering the types of combination, whether Metal-on-Metal, Ceramic-on-Ceramic or Metal-on-Polymer bearing combinations. In addition, other parameters such as the patient's conditions related to their gender, age, race and types of activities they are doing were also taken into new parametric studies. For example, hip joint movement for Western and Asian people is also being studied with the results showing that Asians are more critical to prosthetic impingement that may cause dislocation [63]. There are also studies that show certain activities should be avoided in postoperative THR [64, 65].

2.3.1 Mathematical Formula to Calculate the Range of Motion (ROM)

In the process of defining the safe zone orientation, Fumihiro et al. [56] have introduced a mathematical formula to calculate the ROM for every six basic motions without particularly deciding which ROM is defined as the best criteria. These equations are favored upon this research in which the different safe zone orientation in terms of inclination and anteversion angle could be found. Here, Eq. 2.1 is defined as the pivot on developing the follow-up equation for the ROM.

Fig. 2.6 Schematic diagram
of the prosthetic range of
motion (reproduced from
[56], with permission)

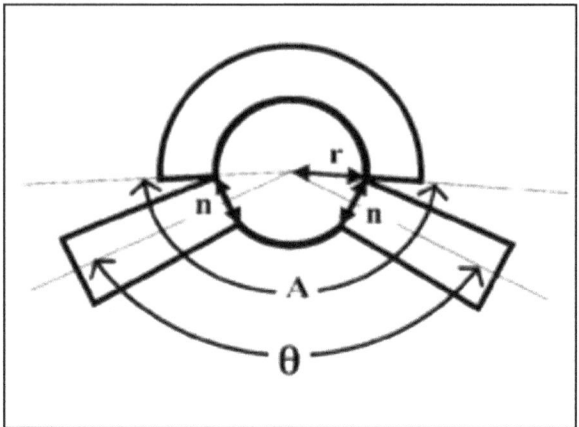

$$\theta = A - 2\sin^{-1}\left(\frac{n/2}{r}\right) = A - 2\sin^{-1}\left(\frac{1}{\text{head/neck ratio}}\right) \qquad (2.1)$$

The (θ) is defined as Oscillation Angle, (A) is the maximum angle of radius movement in the cup, (n) is the neck width at the impingement level and (r) is the radius of the head. A schematic drawing of this equation can be explained as follows (Fig. 2.6).

The follow-up equations that have been developed will be explained with a schematic diagram defining the range of motion. Five factors needed to be understood first in order to get the extreme angles of the basic six ranges of motion. Here, $\theta, \alpha, \beta,$, 'a' and 'b' are the main factors being indicated and all the symbols are derived into the mathematical equation that will determine the angle value of the motions. Oscillation angle (θ) has been mentioned in Eq. 2.1 as it plays a significant role that will influence mathematical equations later. The values of α and β are defined as the inclination angle and anteversion angle, respectively, in these formulas. Meanwhile, the values of 'a' and 'b' are referred to as the angles depending on the design of the femoral stem. On the other hand, E and D that appeared from the equations developed were referred to as constants. The details of the derivations and calculations are available in the Journal of Biomechanics Supplementary Data [56].

For the flexion (FL) and the extension (EXT), the equations are being developed as follows:

$$E_1 = \cos\frac{\theta}{2} - \sin\alpha\,\cos\beta\,\cos a\,\cos b \qquad (2.2)$$

$$D_1 = \left((1 - \sin^2\alpha\,\cos^2\beta)\right)\left(1 - \cos^2 a\,\cos^2 b\right) - E_1^2 \qquad (2.3)$$

Equations 2.2 and 2.3 are the constant equations that were developed by [56]; thus they will be put into the flexion (FL) and extension (EXT) equations later in Eqs. 2.4 and 2.5. There is only a slight difference for the equation developed in flexion (FL) and extension (EXT) cone models where the sign for the numerator is changed from minus to plus sign. Those equations are as follows:

$$FL = \cos^{-1}\left(\frac{((-\sin\beta\cos\alpha\sin b + \cos\alpha\cos\beta\sin a)E_1 - (\cos\alpha\cos\beta\cos a\sin\beta\sin a)\sqrt{D_1})}{(1-\sin^2\alpha\cos^2\beta)(1-\cos^2 a\cos^2 b)}\right) \quad (2.4)$$

$$EXT = \cos^{-1}\left(\frac{((-\sin\beta\cos\alpha\sin b + \cos\alpha\cos\beta\sin a)E_1 + (\cos\alpha\cos\beta\cos a\sin\beta\sin a)\sqrt{D_1})}{(1-\sin^2\alpha\cos^2\beta)(1-\cos^2 a\cos^2 b)}\right) \quad (2.5)$$

The prosthetic ROM cone for the flexion/extension cone can be referred to in Fig. 2.7. Point B is stated as the start point, while $\angle LCB$ is defined as the flexion (FL) cone angle and $\angle MCB$ is defined as the extension (EXT) cone angle.

Meanwhile, internal rotation (IR) and external rotation (ER) equations are as follows with the new constant developed at Eqs. 2.6 and 2.7 that are unaccompanied by other factors. The constant equations are as follows:

$$E_2 = \cos\frac{\theta}{2} - \cos\alpha\cos\beta\sin a \quad (2.6)$$

$$D_2 = \left(1 - \cos^2\alpha\cos^2\beta\right)\cos^2 a - E_2^2 \quad (2.7)$$

From here, the equation for the internal rotation (IR) and external rotation (ER) can be simplified as follows:

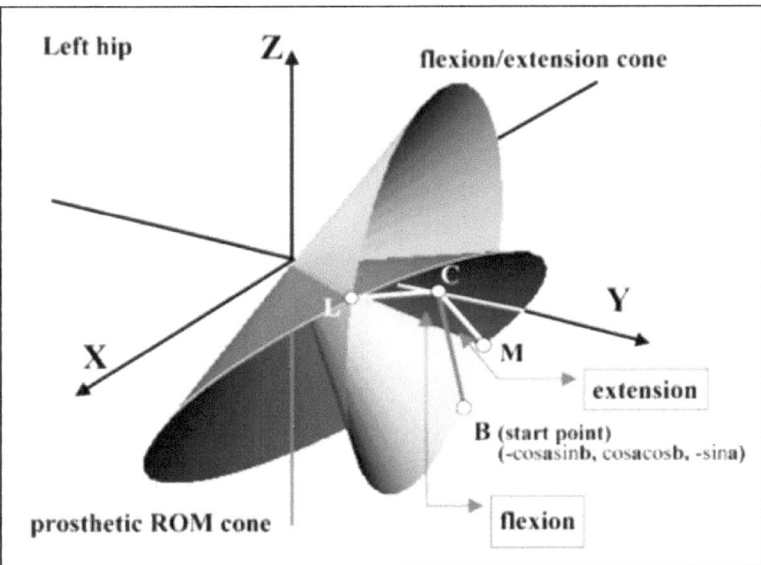

Fig. 2.7 Left THR, the prosthetic ROM cone for flexion and extension. L and M represent the impingement point for flexion and extension, respectively (reproduced from [56], with permission)

$$IR = \cos^{-1}\left(\frac{((-\sin\beta\sin b + \sin\alpha\cos\beta\cos b)E_2 + (\sin\alpha\cos\beta\sin b + \sin\beta\cos b)\sqrt{D_1})}{\cos a(1 - \cos^2\alpha\cos^2\beta)}\right) \qquad (2.8)$$

$$ER = \cos^{-1}\left(\frac{((-\sin\beta\sin b + \sin\alpha\cos\beta\cos b)E_2 - (\sin\alpha\cos\beta\sin b + \sin\beta\,csb)\sqrt{D_1})}{\cos a(1 - \cos^2\alpha\cos^2\beta)}\right) \qquad (2.9)$$

The cone angle of the internal rotation (IR) and external rotation (ER) is shown in Fig. 2.8 taking the schematic drawing of the left hip. $\angle IRDB$ and $\angle ERDB$ are defined as the angles of internal rotation (IR) and external rotation (ER), respectively.

For the case of abduction (ABD) and adduction (ADD), the equations developed are as follows, with also a new constant defined in Eqs. 2.10 and 2.11 as E_3 and D_3 for ease of reference to their respective ABD and ADD constants.

$$E_3 = \cos\frac{\theta}{2} + \sin\beta\cos a\sin b \qquad (2.10)$$

$$D_3 = \left(1 - \cos^2 a\,\sin^2 b\right)\cos^2\beta - E_3^2 \qquad (2.11)$$

Thus, these constant values of E_3 and D_3 can be substituted into the Equations of ABD and ADD of 2.12 and 2.13. Schematic diagram is also included in these parameter sets which will show the rough picture of the angle that defines those angles required upon THR (Fig. 2.9).

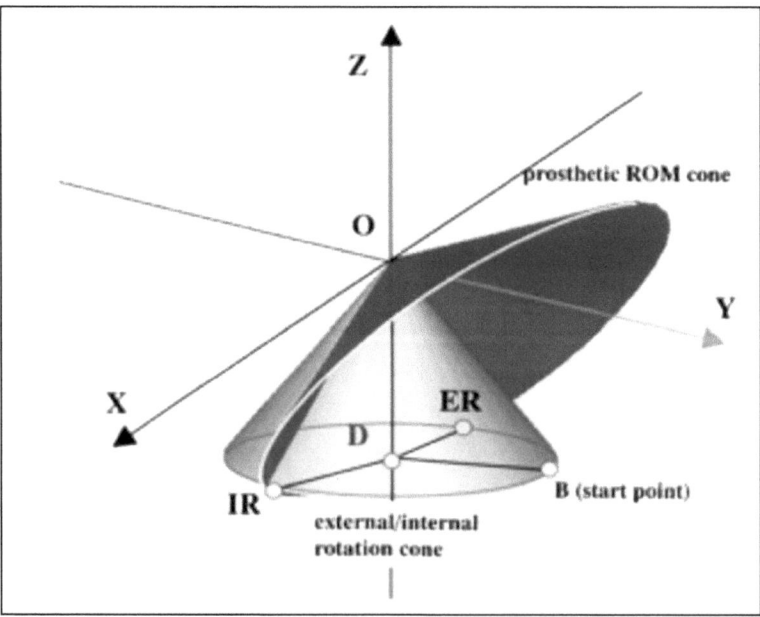

Fig. 2.8 Left THR, the prosthetic ROM cone and internal/external cone (reproduced from [56], with permission)

$$ABD = \cos^{-1}\left(\frac{(\sin\alpha\cos a\cos b + \cos\alpha\sin a)E_3 + (\cos\alpha\cos a\cos b - \sin\alpha\sin a)\sqrt{D_3}}{\cos\beta(1 - \cos^2 a\sin^2 b)}\right) \qquad (2.12)$$

$$ADD = \cos^{-1}\left(\frac{(\sin\alpha\cos a\cos b + \cos\alpha\sin a)E_3 - (\cos\alpha\cos a\cos b - \sin\alpha\sin a)\sqrt{D_3}}{\cos\beta(1 - \cos^2 a\sin^2 b)}\right) \qquad (2.13)$$

These cones show the estimated angle that defines the abduction angle and adduction angle required for the left THR. Fumihiro believes that these equations will simplify the way of reading the best safe zone orientation upon THR at the six basic motions.

Two years later, intended ROM was stipulated by another researcher named Widmer, with flexion (FL) of 130°, extension (EXT) of 40°, abduction (ABD) of 50°, adduction (ADD) of 50°, external rotation (ER) of 40° and internal rotation (IR) of 80° being used as measurement parameters to determine the appropriate cup inclination angle and anteversion angle in his 3D work model [38]. In the case of increasing the intended ROM, the safe zone becomes smaller, and the optimum orientations are shown in Fig. 2.10.

In the next year, a set of new parameters of intended ROM was introduced by the individual who created the mathematical formula in which the parameters are slightly different from Widmer methodology [52]. These new parameters were designated as the set criteria that will allow the finding of appropriate acetabular cup angle which will be run into the mathematical equations developed before. Hence, the following 4 ROM conditions: (1) flexion (FL) at more than 110°, (2) internal rotation at 90°

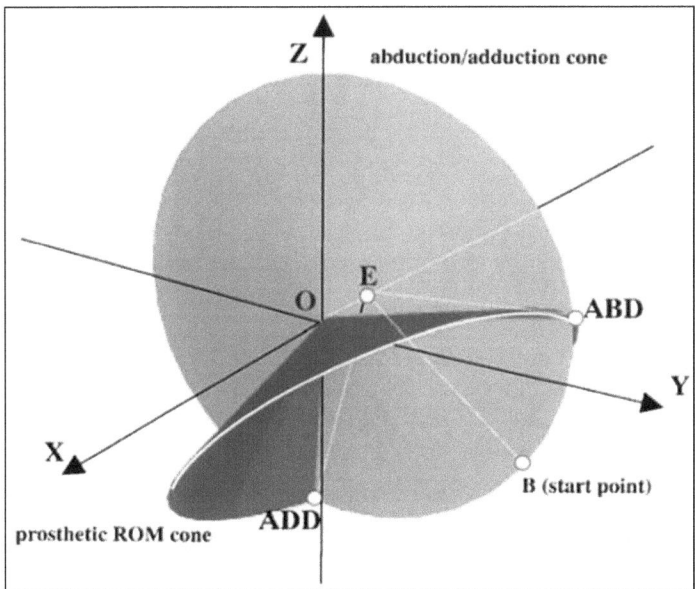

Fig. 2.9 Left THR, the prosthetic ROM cone for abduction and adduction angle (reproduced from [56], with permission)

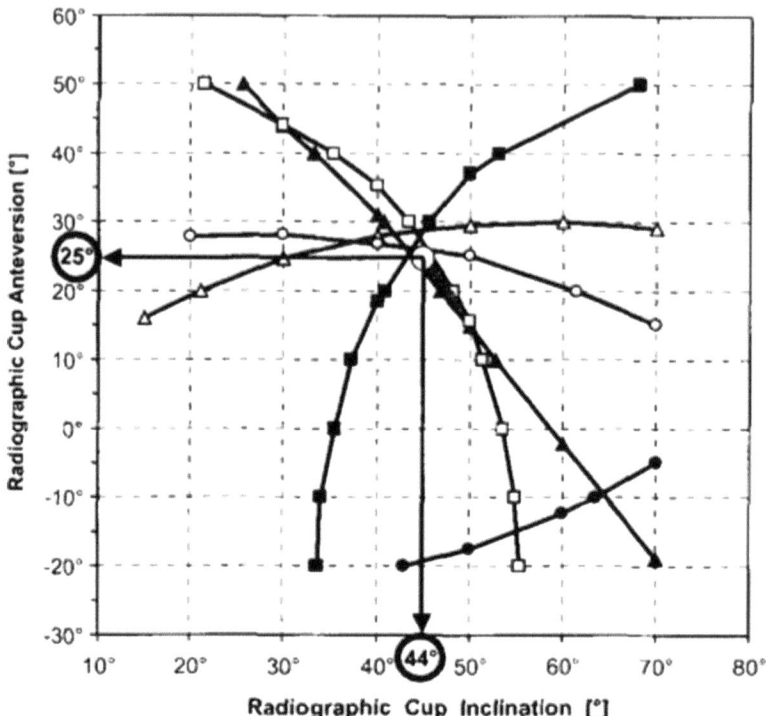

Fig. 2.10 By increasing the intended ROM, the area compliant with safe zone orientation will decrease. In this case, the optimum angle is 44° inclination angles with 25° anteversion angles (reproduced from [38], with permission)

flexion (IRfl90) at more than 30°, (3) external rotation at neutral (ER) at more than 40° and (4) extension (EXT) at more than 30° were arbitrarily selected as the moderate criteria as mentioned in [41]. Meanwhile in 2012, in order to complement the new design parameters, Patel et al. [6] introduced another parameter that they intended to comply with their design, which includes flexion (FL) of 122°, extension (EXT) of 17°, abduction (ABD) of 43°, adduction (ADD) of 40°, internal rotation (IR) of 63° and external rotation (ER) of 30°. However, the author did not specify the rationale behind selecting the six fundamental motions; rather, the author asserted that the increased angle was a more suitable objective for his new implant design. The ranges of motion relationship with the femoral tilt were also being studied with new other parameters in the same year with the flexion (FL) of 130°, extension (EXT) 40°, abduction (ABD) 50°, adduction (ADD) 50°, internal rotation (IR) 80°, external rotation (ER) 40°, internal rotation at 90° flexion (IRfl90) 45° and external rotation at 90° flexion (ERfl90) at 55° in which the study output concluded that femoral tilt is associated with wider range of motion [66].

Previously, D'Lima et al. [67] investigated the effect of the positions of acetabular and femoral components on various head-neck ratios concerning impingement and

ROM, but they did not give an appropriate combination of acetabular and femoral anteversion [41].

References

1. National Joint Registry for England Wales and Northern Ireland, in *13th Annual Report* (2016)
2. AOA, Australian Orthopaedic Association National Joint Replacement Registry, in *Annual Report* (2015)
3. T. Zhang, N.M. Harrison, P.F. McDonnell, P.E. McHugh, S.B. Leen, A finite element methodology for wear–fatigue analysis for modular hip implants. Tribol. Int. **65**, 113–127 (2013)
4. S. Pramanik, A.K. Agarwal, K.N. Rai, Chronology of total hip joint replacement and materials development. Trends Biomater. Artif. Organs **19**(1), 15–26 (2005)
5. T. Sato, N. Sato, Clinical relevance of the hip joint: Part II–Importance of joint distraction. Int. Musculoskelet. Med. **37**(4), 141–145 (2016)
6. D. Patel, T. Goswami, Influence of design parameters on cup-stem orientations for impingement free RoM in hip implants. Med. Eng. Phys. **34**(5), 573–578 (2012)
7. N. Nuño, M. Amabili, Modelling debonded stem–cement interface for hip implants: effect of residual stresses. Clin. Biomech. **17**(1), 41–48 (2002)
8. G.G. Klingenstein, A.M. Yeager, J.D. Lipman, G.H. Westrich, Computerized range of motion analysis following dual mobility total hip arthroplasty, traditional total hip arthroplasty, and hip resurfacing. J. Arthroplasty **28**(7), 1173–1176 (2013)
9. S.M. Kurtz, *UHMWPE Biomaterials Handbook*, 2 Ed. Elsevier (2009)
10. S.E. McElwain, T.A. Blanchet, L.S. Schadler, W.G. Sawyer, Effect of particle size on the wear resistance of alumina-filled PTFE micro- and nanocomposites. Tribol. Trans. **51**, 247–253 (2008)
11. U. Holzwarth, G. Cotogno, *Total Hip Arthroplasty: State of the Art, Challenges and Prospects*, vol. 1 (2012)
12. A. Grazioli, E.T.H. Ek, H.A. Rüdiger, Biomechanical concept and clinical outcome of dual mobility cups. Int. Orthop. **36**(12), 2411–2418 (2012)
13. Y. Xue, X. Liu, J. Sun, PU/PTFE-stimulated monocyte-derived soluble factors induced inflammatory activation in endothelial cells. Toxicol. In Vitro **24**(2), 404–410 (2010)
14. G. Lewis, Properties of crosslinked ultra-high-molecular-weight polyethylene. Biomaterials **22**(4), 371–401 (2001)
15. A.R. Shinge, S.S. Anasane, E.N. Aitavade, S.S. Mahadik, P.V. Mulik, Finite element analysis of modified hip prosthesis. Int. J. Adv. Biotechnol. Res. **2**(2), 278–285 (2011)
16. S. Ramakrishna, J. Mayer, E. Wintermantel, K.W. Leong, Biomedical applications of polymer-composite materials: a review. Compos. Sci. Technol. **61**(9), 1189–1224 (2001)
17. S.G. Ghalme, A. Mankar, Y. Bhalerao, Biomaterials in hip joint replacement. Int. J. Mater. Sci. Eng. **4**(2), 113–125 (2016)
18. J.R.T. Jeffers, M. Browne, A.B. Lennon, P.J. Prendergast, M. Taylor, Cement mantle fatigue failure in total hip replacement : experimental and computational testing. J. Biomech. **40**, 1525–1533 (2007)
19. V. Shim, J. Boheme, C. Josten, I. Anderson, *Use of Polyurethane Foam in Orthopaedic Biomechanical Experimentation and Simulation* (2012)
20. A. Wang, A. Essner, Three-body wear of UHMWPE acetabular cups by PMMA particles against CoCr, alumina and zirconia heads in a hip joint simulator. Wear **250**(1–12), 212–216 (2001)
21. D.D.R. Naudie, D.J. Ammeen, G.A. Engh, C.H. Rorabeck, Wear and osteolysis around total knee arthroplasty. J. Am. Acad. Orthop. Surg. **15**(1), 53–64 (2007)

22. M. Epstein, I. Emri, P. Hartemann, P. Hoet, N. Leitgeb, L.M. Martinez, A. Proykova, L. Rizzo, E. Rodriguez-Farre, L. Rushton, K. Rydzynski, T. Samaras, E. Testai, T. Vermeire, *Scientific Committee on Emerging and Newly Identified Health Risks SCENIHR Opinion on The safety of Metal-on-Metal joint replacements with a particular focus on hip implants* (2014)
23. OECD, Health at a glance 2011: OECD Indicators. OECD Publishing (2011)
24. M.S. Lehil, K.J. Bozic, Trends in total hip arthroplasty implant utilization in the United States. J. Arthroplasty **29**(10), 1915–1918 (2014)
25. S. Shuib, B. Sahari, A. Ahmed Shokri, C.S. Chai, The design improvement of hip implant for total hip replacement (THR). J. Kejuruter. **20**(1), 107–113 (2008)
26. S. Li, B. Huang, Y. Chen, H. Gao, Q. Fan, J. Zhao, W. Su, Hydroxyapatite-coated femoral stems in primary total hip arthroplasty: a meta-analysis of randomized controlled trials. Int. J. Surg. (Lond., Engl.) **11**(6), 477–482 (2013)
27. J.S. Siopack, H.E. Jergesen, Total hip arthroplasty. West. J. Med. **162**(3), 243–249 (1995)
28. J.J. Prokopetz, E. Losina, R.L. Bliss, J. Wright, J.A. Baron, J.N. Katz, Risk factors for revision of primary total hip arthroplasty: a systematic review. BMC Musculoskelet. Disord. **13**, 251 (2012)
29. M. Saam, J.B. Kevin, D.R. Michael, M. Henrik, M.C.J. John, Comparison of cemented and uncemented fixation in total knee arthroplasty. Orthopedics **78**(3), 315–326 (2007)
30. M. Pennington, R. Grieve, J.S. Sekhon, P. Gregg, N. Black, J.H. van der Meulen, Cemented, cementless, and hybrid prostheses for total hip replacement: cost effectiveness analysis. BMJ (Clin. Res. Ed.) **346**, 1–14 (2013)
31. C.C. John, H.H. William, Primary hybrid total hip replacement, performed with insertion of the acetabular component without cement and a precoat femoral component with cement. An average ten-year follow-up study. J. Bone Joint Surg. Am. **81**(2), 247–255 (1999)
32. R.J. Kane, W. Yue, J.J. Mason, R.K. Roeder, Improved fatigue life of acrylic bone cements reinforced with zirconia fibers. J. Mech. Behav. Biomed. Mater. **3**(7), 504–511 (2010)
33. B.A.O. Mccormack, C.D. Walsh, S.P. Wilson, P.J. Prendergast, A statistical analysis of micro-crack accumulation in PMMA under fatigue loading: applications to orthopaedic implant fixation. Int. J. Fatigue **20**(3), 581–593 (1998)
34. B.R. Burroughs, B. Hallstrom, G.J. Golladay, D. Hoeffel, W.H. Harris, Range of motion and stability in total hip arthroplasty with 28-, 32-, 38-, and 44-mm femoral head sizes: an in vitro study. J. Arthroplasty **20**(1), 11–19 (2005)
35. J. Girard, Femoral head diameter considerations for primary total hip arthroplasty. Orthop. Traumatol. Surg. Res. **101**(1), 25–29 (2015)
36. D.R. Pedersen, J.J. Callaghan, T.L. Johnston, G.B. Fetzer, R.C. Johnston, Comparison of femoral head penetration rates between cementless acetabular components with 22-mm and 28-mm heads. J. Arthroplasty **16**(Suppl 1), 111–115 (2001)
37. A. Matsushita, Y. Nakashima, S. Jingushi, T. Yamamoto, A. Kuraoka, Y. Iwamoto, Effects of the femoral offset and the head size on the safe range of motion in total hip arthroplasty. J. Arthroplast. **24**(4), 646–651 (2009)
38. K.-H. Widmer, B. Zurfluh, Compliant positioning of total hip components for optimal range of motion. J. Orthop. Res.: Off. Publ. Orthop. Res. Soc. **22**(4), 815–821 (2004)
39. X. Hua, B.M. Wroblewski, Z. Jin, L. Wang, The effect of cup inclination and wear on the contact mechanics and cement fixation for ultra high molecular weight polyethylene total hip replacements. Med. Eng. Phys. **34**(3), 318–325 (2012)
40. W.L. Walter, G.M. Insley, W.K. Walter, M.A. Tuke, Edge loading in third generation alumina ceramic-on-ceramic bearings: stripe wear. J. Arthroplast. **19**(4), 402–413 (2004)
41. F. Yoshimine, The safe-zones for combined cup and neck anteversions that fulfill the essential range of motion and their optimum combination in total hip replacements. J. Biomech. **39**(7), 1315–1323 (2006)
42. M. Alvarez-Vera, G.R. Contreras-Hernandez, S. Affatato, M.A.L. Hernandez-Rodriguez, A novel total hip resurfacing design with improved range of motion and edge-load contact stress. Mater. Des. **55**, 690–698 (2014)

43. Y.-S. Yoon, A.J. Hodgson, J. Tonetti, B.A Masri, C.P. Duncan, Resolving inconsistencies in defining the target orientation for the acetabular cup angles in total hip arthroplasty. Clin. Biomech. (Bristol, Avon) **23**(3), 253–259, (2008)

44. C.L. Harrison, A.I. Thomson, S. Cutts, P.J. Rowe, P.E. Riches, Research synthesis of recommended acetabular cup orientations for total hip arthroplasty. J. Arthroplast. **29**(2), 377–382 (2014)

45. A.S. Mclawhorn, P.K. Sculco, K.D. Weeks, D. Nam, D.J. Mayman, Targeting a new safe zone: a step in the development of patient-specific component positioning for total hip arthroplasty. Am. J. Orthop. **44**(6), 270–276 (2015)

46. G.E. Lewinnek, J.L. Lewis, R. Tarr, C.L. Compere, J. Zimmerman, Dislocations after total hip replacement arthroplasties. J. Bone Joint Surg. Am. (60), 217–220 (1978)

47. D.E. McCollum, W.J. Gray, Dislocation after total hip arthroplasty causes and prevention. Clin. Orthop. Relat. Res. (261), 159–170 (1990)

48. A. Malik, A. Maheshwari, L.D. Dorr, Impingement with total hip replacement. J. Bone Jt. Surg. Am. **89**(8), 1832–1842 (2007)

49. M. Seki, N. Yuasa, K. Ohkuni, Analysis of optimal range of socket orientations in total hip arthroplasty with use of computer- aided design simulation. J. Orthop. Res. **16**(4), 513–517 (1998)

50. B.M. Jolles, P. Zangger, P.-F. Leyvraz, Factors predisposing to dislocation after primary total hip arthroplasty: a multivariate analysis. J. Arthroplast. **17**(3), 282–288 (2002)

51. K.-H. Widmer, A simplified method to determine acetabular cup anteversion from plain radiographs. J. Arthroplast. **19**(3), 387–390 (2004)

52. F. Yoshimine, The influence of the oscillation angle and the neck anteversion of the prosthesis on the cup safe-zone that fulfills the criteria for range of motion in total hip replacements. The required oscillation angle for an acceptable cup safe-zone. J. Biomech. **38**(1), 125–132 (2005)

53. R. Biedermann, A. Tonin, M. Krismer, F. Rachbauer, G. Eibl, B. Stöckl, Reducing the risk of dislocation after total hip arthroplasty. J. Bone Jt. Surg., Br. **87**(6), 762–769 (2005)

54. F.J. Kummer, S. Shah, S. Iyer, P.E. DiCesare, The effect of acetabular cup orientations on limiting hip rotation. J. Arthroplast. **14**(4), 509–513 (1999)

55. D. Kluess, H. Martin, W. Mittelmeier, K.-P. Schmitz, R. Bader, Influence of femoral head size on impingement, dislocation and stress distribution in total hip replacement. Med. Eng. Phys. **29**(4), 465–471 (2007)

56. F. Yoshimine, K. Ginbayashi, A mathematical formula to calculate the theoretical range of motion for total hip replacement. J. Biomech. **35**(7), 989–993 (2002)

57. M.C. Callanan, B. Jarrett, C.R. Bragdon, D. Zurakowski, H.E. Rubash, A.A Freiberg, H. Malchau, The John Charnley Award: risk factors for cup malpositioning: quality improvement through a joint registry at a tertiary hospital. Clin. Orthop. Relat. Res. **469**(2), 319–329 (2011)

58. N.J. Little, C.A Busch, J.A Gallagher, C.H. Rorabeck, R.B. Bourne, Acetabular polyethylene wear and acetabular inclination and femoral offset. Clin. Orthop. Relat. Res. **467**(11), 2895–2900 (2009)

59. H. Miki, T. Kyo, Y. Kuroda, I. Nakahara, N. Sugano, Risk of edge-loading and prosthesis impingement due to posterior pelvic tilting after total hip arthroplasty. Clin. Biomech. (Bristol, Avon) **29**(6), 607–613 (2014)

60. K.-H. Widmer, M. Majewski, The impact of the CCD-angle on range of motion and cup positioning in total hip arthroplasty. Clin. Biomech. (Bristol, Avon) **20**(7), 723–728 (2005)

61. W.-T. Ji, K. Tao, C.-T. Wang, A three-dimensional parameterized and visually kinematic simulation module for the theoretical range of motion of total hip arthroplasty. Clin. Biomech. (Bristol, Avon) **25**(5), 427–432 (2010)

62. G. Saxler, A. Marx, D. Vandevelde, U. Langlotz, M. Tannast, M. Wiese, U. Michaelis, G. Kemper, P.A. Grützner, R. Steffen, M. von Knoch, T. Holland-Letz, K. Bernsmann, The accuracy of free-hand cup positioning—a CT based measurement of cup placement in 105 total hip arthroplasties. Int. Orthop. **28**(4), 198–201 (2004)

63. E. Saputra, I. Budiwan, R. Ismail, J. Jamari, E. Van Der Heide, Jurnal Teknologi Full paper Numerical Simulation of Artificical Hip Joint Movement for Western and. Jurnal Teknologi **66**(3), 53–58 (2014)
64. X. Hua, J. Li, Z. Jin, J. Fisher, The contact mechanics and occurrence of edge loading in modular metal-on-polyethylene total hip replacement during daily activities. Med. Eng. Phys. **38**(6), 518–525 (2016)
65. S. Shuib, B. Sahari, W. Voon, A.H. Kadarman, Stress analysis of femoral bone resorption with implant. Trends Biomater. Artif. Organs **27**(2), 88–92 (2013)
66. T. Renkawitz, M. Haimerl, L. Dohmen, S. Gneiting, P. Lechler, M. Woerner, H.-R. Springorum, M. Weber, P. Sussmann, E. Sendtner, J. Grifka, The association between Femoral Tilt and impingement-free range-of-motion in total hip arthroplasty. BMC Musculoskelet. Disord. **13**(1), 65 (2012)
67. D. D'Lima, A. Urguhart, K. Buehler, R. Walker, C. Colwell, The effect of the orientation of the acetabular and femoral components on the range of motion of the hip at different head-neck ratios. J. Bone Jt. Surg Am. **82**(3), 315–321 (2000)

Chapter 3
Finite Element Analysis (FEA) of Metal-On-Polymer (MoP) and Acetabular Cup Material

3.1 Finite Element (FE) Studies of MoP THR

Finite element method is a powerful tool used to analyze every aspect related to the engineering field. In THR, finite element studies are utilized to find the desired results that aid in improving the THR implant. Although there are many FEA studies done throughout the introduction of THR, this subchapter will be focused on MoP THR that is related to this research.

3.1.1 The Parameters Consideration of MoP FEA Analysis

The review from previous work on THR shows that there are many considerations to analyze the THR by using finite element. Table 3.1 shows the summarization of FEA methodology important parameters upon MoP THR.

3.1.2 FEA Analysis Results of Metal-On-Polymer THR

The results of the FEA conducted previously upon MoP THR are briefly explained in this subsection. In 1999, a novel design of MoP acetabular components was introduced with the aim of resisting dislocation issue [1]. The results show that new curved lip design of acetabular cup will reduce about 50% Von-Mises stress and 28% more resistance to dislocation. The curved lip design was indicated as the best design for the MoP THR compared to other design proposal. In 2005, four new designs of cemented MoP were introduced with three inclination angles and a 3000N vertical loading condition, resulting in changes to the design, thickness, orientation and clearance that affect the mechanical stress experienced by UHMWPE acetabular cup [2].

M. F. b. A. Manap et al., *Total Hip Replacement (THR)*,
SpringerBriefs in Applied Sciences and Technology,
https://doi.org/10.1007/978-981-96-0975-8_3

Table 3.1 The summarization of FEA methodology upon *MoP THR*

References	Software	Loading conditions (N)	Material properties (MPa)			Constrained	Coefficient of friction (μ)	Number of elements	Number of nodes
			Acetabular cup	Shell	Femoral head				
Scifert et al. [1]	ABAQUS and Patran	942	634.97	110,000	Rigid	Shell	0.5	360 ~ 3200	N/A
Korhonen et al. [2]	ABAQUS	3000	690	2740	193,000	Shell	N/A	15,834 ~ 16,072	N/A
Kluess et al. [3]	ABAQUS and Patran	$F_r = 506$; $F_f = 1288$	945	Rigid	Rigid	Rotational DOF of acetabular cup	N/A	10,800	N/A
Ghaffari et al. [4]	ABAQUS	Seven conditions	945	Excluded	Rigid	Acetabular cup	N/A	N/A	N/A
Hua et al. [5]	ABAQUS	2500	850	2500	Rigid	Pelvis model	0.1	± 12,000	± 10,000
Shankar et al. [6]	ANSYS	2500	2200	Various	Various	Shell	N/A	60,800	N/A
Saputra et al. [7]	ABAQUS	$F_x = 15$; $F_y = 270$; $F_z = -427.5$	945	Excluded	Rigid	All DOF of acetabular cup	N/A	9000	N/A
Hua et al. [8]	ABAQUS	2500	1000	116,000	Rigid	Pelvis model	0.083 and 0.15	± 92,000	N/A
Korduba et al. [9]	ANSYS	2450	945	Rigid	Rigid	Shell	Frictionless	N/A	N/A
Zietz et al. [10]	ABAQUS	Four conditions	N/A	N/A	N/A	Shell	0.16 and 0.06	74,000	114,000
Wang et al. [11]	ABAQUS and Hypermesh	Five conditions	875	2500	220,000	Pelvis model	Frictionless	± 500,000	N/A
Hua et al. [12]	ABAQUS	Six conditions	1000	116,000	Rigid	Pelvis model	0.083 and 0.15	± 92,000	N/A

At this timeframe, the focus is more on designing new acetabular components that will perform better based on FEA results analysis.

However, two years later in 2007, a FEA study was conducted to find the effect of femoral head size with different orientation and the results show that complying with the range of motion should be the goal for any new implant design [3]. In 2012, a study using only 24 mm femoral head was performed with more orientation combination and various loading conditions that show the localized stress at impingement site will damage the acetabular cup [4]. At the same year too, another researcher named Hua et al. focused on cemented MoP (22.59mm femoral head) with 2500 N loading condition that exhibited cup inclination increase of up to 65° and had resulted in huge increment of Von-Mises stress at bone cement interface [5]. In 2013, a research study conducted shows that among different combination of acetabular component, the MoP is regarded as the better combination as the MoP reduces Von-Mises and contact pressure of about 90% and 91%, respectively [6]. There are also studies that used Hertzian Contact Theory to support the FEA method on the objective to improve the THR implant [13, 14]. In the same year, more studies related to the orientation of acetabular cup were conducted using 28mm femoral head at 30°, 45° and 60° inclination angles with 16 loading conditions which show that 45° inclination presents lower linear wear but the highest volumetric wear [15]. In addition to that, a study compared the Western and Japanese style activities using FEA in which the data shows that impingement was found in Western picking-up activities and critical range of motion was observed in Seiza Japanese activities [7].

In 2014, a researcher conducted a new study objective of edge-loading effect using 36mm femoral head diameter and micro separation techniques. The FEA was performed with pelvis model included in the study with more than 92,000 elements and three loading conditions. The study shows that lateral micro separation will cause an increase in Von-Mises and contact pressure; thus correct positioning is important to avoid edge-loading effect [16]. During this time frame, the aim of doing FEA focused on other parameter conditions rather than on the new design of the acetabular components. The preference software, loading condition, meshing, material properties and boundary condition are different upon every researcher, yet FEA proves to be important to conduct studies in MoP THR.

Based on these reviews, most of the experimental works will be supported by FEA method and vice versa. In 2014, another researcher conducted an experiment using Hip Simulator Testing and FEA method was run in order to assess the contact pressure and contact area with different orientation angles [9]. The test conducted shows strong correlation between acetabular cup inclination and volumetric polyethylene wear. Later the next year, two new studies using Hip Simulator Testing of MoP THR were done, with the FEA results used as the validation method based on experimental studies. One study chose the XLPE material to conduct the wear study, supported by FEA analysis results [10]. Meanwhile, the other study, conducted by using UHMWPE material, found that increasing the progressive wear will increase the contact pressure and Von-Mises stress [17]. In 2016, FEA was conducted to study the edge-loading effect during daily activities, which showed that edge-loading was predicted during normal walking, ascending and descending stairs activities at more than a 55° inclination angle.

The MoP THR finite element related to orientation was conducted by [3, 9, 12] in their work, which uses various set of orientation disregarding the safe zone claim. This research aims to add more parameter studies, such as optimum safe zone orientation, into the FEA analysis; thus the data acquired will be more accurate. In conclusion, the intention to improve the implant performance of MoP THR in terms of design, orientation, size and other parameters could be achieved by using FEA although further studies need to be conducted later.

3.2 Acetabular Cup Materials

The issues exposed by the acetabular cup 'insert/liner' material are agreable with this research question; among them are the dislocation, fracture, indentation deformation, inflammation due to wear and micro-motion on the bone cement fixation [3, 18]. The improvements and new fabrications were conducted to reinforce the material's mechanical properties, as performed by [19, 20] in their works. Most of the studies resulted in significant improvements in the cement and the acetabular cup properties in terms of mechanical wear, stiffness and toughness. Implant–cement fixation is generally achieved either by selecting an implant surface texture that creates a mechanical interlock with the bone cement or by an implant with geometry that maintains stability such as polished tapered stems [21] with PMMA common to it. Even though PMMA is an unarguably suitable material, the present pores from the cement itself create a crack riser [22] that eventually leads to failure. Due to human activity behavior, femoral stem and bone cement debonding may occur based on stress shielding that causes friction and eventually, crack propagation starts to appear [23]. Based on trends, most of the implants have evolved into cementless hip implant as porous femoral stems and acetabular cups (metal backing) increase the rate of bone ingrowths into the components [24]. The metal backing will be used as the replacement of the PMMA and attached together with acetabular liner cup. On the other side, cementless implant exposes greater demerits to older generations compared to the young generation, as the PMMA cement can act as compensation for bone defects [25]. Thus, for younger patients who have more active bone tissue, the cementless THR is more preferred as their replacement.

Aside from the cemented acetabular component, UHMWPE is regarded as the best solution for the polymer parts in THR due to its mechanical properties [26]. Since revolutionized by Sir John Charnley more than half a century ago, a system that commonly uses stainless steel femoral head and UHMWPE acetabular cup has become a standard use in THR [27]. The purpose of acetabular components made of polyethylene in hip prostheses mainly includes shock protection/ absorption, and displacement adjustments during gait motion, including mal-alignments and rotational strains [28]. Even though UHMWPE is highly considered a successful polymer, the articulation between metal and polymer commits severe wear debris to the implant parts that is dangerous to the human body. Aseptic loosening is common when wear debris from UHMWPE occurs, leading to a revision of that implant [29]. Aseptic loosening is less-frequent in MoM or CoC prostheses compared to MoP since wear volume data on MoM and CoC are limited [27]. For the past decade, MoM prostheses have become the preferred hip implant due to the design that mimics the original hip joint [25]. Increasing the usage of a bigger femoral head has been discussed with the effect that it may reduce the dislocation and increase the range of motion of the ball socket joints. In contrast, a bigger femoral head may potentially give disadvantage with more surface motion in the area, thus increasing the wear debris from the hip implant. There is another issue faced upon in which PMMA eventually embedded in between the articulation of UHMWPE acetabular cup and femoral head after some time. Based on research, many scenarios proved that PMMA affects polyethylene wear as it may increase the wear rate indirectly [30].

For most cementless THR, elimination of PMMA usage is common with a metal backing used as a substitute. It is important to note that materials are being used in the human body as shown in Table 4 [31] and the application of these materials is made by considering many aspects and the contribution of the materials in the human body. Specifically, the data on the joint replacement bone shows that enormous material selection has been tried out with the aim of improving the hip implant life (Table 3.2).

Among various materials incorporated into the UHMWPE, one of the highlights for the usage of UHMWPE is the process of incorporating UHMWPE with natural coral (NC), which exhibits micro-hardness and scratch resistance that keep increasing with higher filler loading of NC particle [19]. Later in 2011, Chang et al. [32] investigated the effect of incorporating zinc oxide (ZnO) into the UHMWPE and the data shows that the performance in terms of tensile strength decreased with higher filler loading compared to pure UHMWPE. Two years later in 2013, he again proposed the Zeolite-reinforced UHMWPE for implant application which shows that the composites were fully covered, smooth and continuous compared to pure UHMWPE which will perform better at the articulate surfaces between femoral head and acetabular cup [20].

UHMWPE is considered a suitable material for the acetabular cup in THR with its superior properties. However, the wear debris generated still remains an issue, attacking the immune system of one's body, which is a challenge that remains open for

Table 3.2 Among the application of composite implant used in human body [31]

Applications	Types of materials
Joint replacements	PET/PHEMA, KF/PMA, KF/PE, CF/PTFE, CF/PLLA, GF/PU, PET/PU, PTFE/PU, CF/PTFE, CF/C, CF/UHMWPE, UHMWPE, CF/Epoxy, CF/PS, CF/PEEK, CF/UHMWPE, CF/PE
Bone cement	Bone particles/PMMA, Titanium/PMMA, UHMWPE/PMMA, GF/PMMA, CF/PMMA, Bio-Glass/Bis- GMA
Bone replacement materials	HA/PHB, HA/PEG-PHB, CF/PTFE, PET/PU, HA/HDPE, HA/PE, Bio- Glass/PE, Bio-Glass/PHB, Bio-Glass/PS, HA/PLA

future improvement. Osteolysis causes the implant loosening and ultimately requires revision [33]. Indentation also appears on the superior region of the acetabular cup when dealing with smaller femoral head diameter [3]. Besides, the analysis that has been done with the Metal-on-Polymer (MoP) combinations, along with overheating of the two articulating surfaces, also lead to the failure of the cup due to insufficient lubrication at the contact region [34].

Before, there were many types of research focusing on the failure of the acetabular cup. Abrasion wear, dislocation, fatigue failure, delamination, surface crack and manufacturing defects are among the most contributors to acetabular cup failure. For cases based on hard-on-soft combinations, the wear generated and contact stress depending on cup orientation were studied and the results show that correct positioning of the acetabular component to avoid edge-loading is a proven important characteristic in THR [5, 16].

3.2.1   Epoxy-UHMWPE Materials

Although Epoxy has been studied for joint replacement materials previously, the review conducted shows that Epoxy-UHMWPE composite combinations have never been tested for the THR application. However, there is a study that shows that the usage of Epoxy resin to the UHMWPE will serve as a physical cross-linking agent to limit the motion of PE molecules, consequently improving the mechanical properties [35]. The preparation of the samples was conducted using hand-layup techniques with three stages of post-curing heating to improve the homogeneity of UHMWPE powder. Another study of an Epoxy-UHMWPE composite was also conducted, with the results showing that using nano-Epoxy instead of pure Epoxy improves the wettability of UHMWPE [36]. The samples were prepared by using a digital sonifier of the sonification method. In 2015, another study of Epoxy-UHMWPE composite was conducted which showed that the post-curing temperature gives effect to the composite stiffness and modulus [37]. The samples were prepared using the vacuum-assisted resin transfer molding (VaRTM) method which was almost similar to hand-layup techniques but with vacuum modification. The previous studies of Epoxy-UHMWPE exhibit that adding Epoxy to the UHMWPE filler/fiber will improve the mechanical properties of the composite.

It is believed that both the criteria of acetabular orientation and material selection in THR are crucial to be studied; thus understanding both relationships is needed to be discussed in this project. The failure of the hip implant will be studied by taking the acetabulum part components as the objectives; thus aiming to develop the acetabular components orientation and acetabular cup materials to improve the implant life.

3.3 Summary of Chapter Three

The key points are as follows:

- Metal-on-Polymer THR is divided into two types which are cemented and cementless and they are still applicable in the THR application.
- Safe zone orientation is introduced purposely to tackle the issues that arise from the THR procedures in which researchers have different conclusions about the safe zone orientation.
- The optimum safe zone orientation requires ROM to be complied which enables the patient to perform the activities of daily living (ADL).
- FEA is a powerful tool to conduct studies on various parameter conditions related to THR implant.
- MoP THR conducted using FEA related to orientation exhibits that researchers intend to use different sets of orientation disregarding safe zone orientation.
- Acetabular component material selections vary based on clinical studies and much effort is conducted to improve the material properties of THR.
- Previous studies on Epoxy-UHMWPE show that Epoxy may improve the mechanical properties of UHMWPE filler/fiber.

References

1. C.F. Scifert, T.D. Brown, J.D. Lipman, Finite element analysis of a novel design approach to resisting total hip dislocation. Clin. Biomech. (Bristol, Avon) **14**, 697–703 (1999)
2. R.K. Korhonen, A. Koistinen, Y.T. Konttinen, S.S. Santavirta, R. Lappalainen, The effect of geometry and abduction angle on the stresses in cemented UHMWPE acetabular cups—finite element simulations and experimental tests. Biomed. Eng. Online **4**(1), 32 (2005)
3. D. Kluess, H. Martin, W. Mittelmeier, K.-P. Schmitz, R. Bader, Influence of femoral head size on impingement, dislocation and stress distribution in total hip replacement. Med. Eng. Phys. **29**(4), 465–471 (2007)
4. M. Ghaffari, R. Nickmanesh, N. Tamannaee, F. Farahmand, The impingement-dislocation risk of total hip replacement: effects of cup orientation and patient maneuvers, in *Conference Proceedings: IEEE Engineering in Medicine and Biology Society*, vol. 34, no. 1 (2012), pp. 6801–6804
5. X. Hua, B.M. Wroblewski, Z. Jin, L. Wang, The effect of cup inclination and wear on the contact mechanics and cement fixation for ultra high molecular weight polyethylene total hip replacements. Med. Eng. Phys. **34**(3), 318–325 (2012)
6. S. Shankar, L. Prakash, M. Kalayarasan, Finite element analysis of different contact bearing couples for human hip prosthesis. Int. J. Biomed. Eng. Technol. **11**(1), 66–80 (2013)
7. E. Saputra, I. Budiwan, R. Ismail, J. Jamari, E. Van Der Heide, Jurnal teknologi full paper numerical simulation of articifical hip joint movement for western and Japanse-style activities. Jurnal Teknologi **66**(3), 53–58 (2014)
8. X. Hua, L. Wang, M. Al-Hajjar, Z. Jin, R.K. Wilcox, J. Fisher, Experimental validation of finite element modelling of a modular metal-on-polyethylene total hip replacement. Proc. Inst. Mech. Eng. Part H: J. Eng. Med. **228**(7), 682–692 (2014)

9. L.A. Korduba, A. Essner, R. Pivec, P. Lancin, M.A. Mont, A. Wang, R.E. Delanois, Effect of acetabular cup abduction angle on wear of ultrahigh-molecular-weight polyethylene in hip simulator testing. Am. J. Orthop.Orthop. **43**(10), 466–471 (2014)
10. C. Zietz, C. Fabry, F. Baum, R. Bader, D. Kluess, The divergence of wear propagation and stress at steep acetabular cup positions using ceramic heads and sequentially cross-linked polyethylene liners. J. Arthroplast. (2015)
11. L. Wang, X. Liu, D. Li, F. Liu, Z. Jin, Contact mechanics studies of an ellipsoidal contact bearing surface of metal-on-metal hip prostheses under micro-lateralization. Med. Eng. Phys. **36**(4), 419–424 (2014)
12. X. Hua, J. Li, Z. Jin, J. Fisher, The contact mechanics and occurrence of edge loading in modular metal-on-polyethylene total hip replacement during daily activities. Med. Eng. Phys. **38**(6), 518–525 (2016)
13. T. Sato, N. Sato, Clinical relevance of the hip joint: Part II–Importance of joint distraction. Int. Musculoskelet. Med. **37**(4), 141–145 (2016)
14. K.V Nemade, V.K. Tripathi, A mathematical model to calculate contact stresses in artificial human hip joint. Int. J. Eng. Res. Dev. **6**(12), 119–123 (2013)
15. R.D. Queiroz, A.L.L. Oliveira, F.C. Trigo, J.A. Lopes, A finite element method approach to compare the wear of acetabular cups in polyethylene according to their lateral tilt in relation to the coronal plane. Wear **298–299**(1), 8–13 (2013)
16. X. Hua, J. Li, L. Wang, Z. Jin, R. Wilcox, J. Fisher, Contact mechanics of modular metal-on-polyethylene total hip replacement under adverse edge loading conditions. J. Biomech.Biomech. **47**(13), 3303–3309 (2014)
17. L. Wang, W. Yang, X. Peng, D. Li, S. Dong, S. Zhang, J. Zhu, Z. Jin, Effect of progressive wear on the contact mechanics of hip replacements—does the realistic surface profile matter? J. Biomech.Biomech. **48**(6), 1112–1118 (2015)
18. R.J. Kane, W. Yue, J.J. Mason, R.K. Roeder, Improved fatigue life of acrylic bone cements reinforced with zirconia fibers. J. Mech. Behav. Biomed. Mater.Behav. Biomed. Mater. **3**(7), 504–511 (2010)
19. S. Ge, S. Wang, X. Huang, Increasing the wear resistance of UHMWPE acetabular cups by adding natural biocompatible particles. Wear **267**(5–8), 770–776 (2009)
20. B.P. Chang, H.M. Akil, R.M. Nasir, Mechanical and tribological properties of zeolite-reinforced UHMWPE composite for implant application. Procedia Eng. **68**, 88–94 (2013)
21. W.R. Walsh, M.J. Svehla, J. Russell, M. Saito, T. Nakashima, R.M. Gillies, W. Bruce, R. Hori, Cemented fixation with PMMA or Bis-GMA resin hydroxyapatite cement: effect of implant surface roughness. Biomaterials **25**(20), 4929–4934 (2004)
22. N.E. Bishop, S. Ferguson, S. Tepic, Porosity reduction in bone cement at the cement-stem interface. J. Bone Jt. Surg. (Br) **78**(3), 349–356 (1992)
23. J. Geringer, J. Pellier, F. Cleymand, B. Forest, Atomic force microscopy investigations on pits and debris related to fretting-corrosion between 316L SS and PMMA. Wear **292–293**, 207–217 (2012)
24. M.S. Lehil, K.J. Bozic, Trends in total hip arthroplasty implant utilization in the United States. J. ArthroplastyArthroplasty **29**(10), 1915–1918 (2014)
25. U. Holzwarth, G. Cotogno, *Total Hip Arthroplasty: State of the Art, Challenges and Prospects*, vol. 1 (2012)
26. M. Slouf, S. Eklova, J. Kumstatova, S. Berger, H. Synkova, A. Sosna, D. Pokorny, M. Spundova, G. Entlicher, Isolation, characterization and quantification of polyethylene wear debris from periprosthetic tissues around total joint replacements. Wear **262**(9–10), 1171–1181 (2007)
27. J.J. Elsner, Y. Mezape, K. Hakshur, M. Shemesh, E. Linder-Ganz, A. Shterling, N. Eliaz, Wear rate evaluation of a novel polycarbonate-urethane cushion form bearing for artificial hip joints. Acta Biomater. Biomater. **6**(12), 4698–4707 (2010)
28. L. Puppulin, N. Sugano, W. Zhu, G. Pezzotti, Structural modifications induced by compressive plastic deformation in single-step and sequentially irradiated UHMWPE for hip joint components. J. Mech. Behav. Biomed. Mater.Behav. Biomed. Mater. **31**(1), 86–99 (2014)

29. B.T. McMullin, M.-Y. Leung, A.S. Shanbhag, D. McNulty, J.D. Mabrey, C.M. Agrawal, Correlating subjective and objective descriptors of ultra high molecular weight wear particles from total joint prostheses. Biomaterials **27**(5), 752–757 (2006)

30. A. Wang, A. Essner, Three-body wear of UHMWPE acetabular cups by PMMA particles against CoCr, alumina and zirconia heads in a hip joint simulator. Wear **250**(1–12), 212–216 (2001)

31. N. Patel, P. Gohil, A review on biomaterials: scope, applications & human anatomy significance. Int. J. Emerg. Technol. Adv. Eng. **2**(4), 91–101 (2012)

32. B.P. Chang, H.M. Akil, R.M. Nasir, S. Nurdijati, Mechanical and antibacterial properties of treated and untreated zinc oxide filled UHMWPE composites. J. Thermoplast. Compos. Mater.Thermoplast. Compos. Mater. **24**(5), 653–667 (2011)

33. K.S. Katti, Biomaterials in total joint replacement. Colloids Surfaces. B Biointerfaces **39**(3), 133–142 (2004)

34. N.D.L. Burger, P.L. de Vaal, J.P. Meyer, Failure analysis on retrieved ultra high molecular weight polyethylene (UHMWPE) acetabular cups. Eng. Fail. Anal. **14**(7), 1329–1345 (2007)

35. P. Taylor, S. Liu, X. Wang, Y. Wang, Y. Wang, Journal of macromolecular science, part B : physics study on the structure and properties of UHMWPE/Epoxy resin composite fiber study on the structure and properties of UHMWPE/Epoxy resin composite fiber. J. MacroMolecular Sci. **45**(4), 37–41 (2016)

36. S. Neema, A. Salehi-khojin, A. Zhamu, W.H. Zhong, S. Jana, Y.X. Gan, Wettability of nano-epoxies to UHMWPE fibers. J. Colloid Interface Sci. **299**(2), 332–341 (2006)

37. Y. Kang, S. Oh, J.S. Park, Properties of UHMWPE fabric reinforced epoxy composite prepared by vacuum-assisted resin transfer molding. Fibers Polym. **16**(6), 1343–1348 (2015)

Chapter 4
Numerical and Mathematical Modeling for Acetabular Cup Orientation

4.1 Introduction

This chapter describes the procedures required to accomplish the main objectives of the research by developing acetabular cup orientation and new composite system to improve the performance of the implant. Three main procedure stages will be discussed in this chapter with the first stage of finding the safe zone orientation based on a numerical study in MATLAB. The second stage is studying the mechanical properties of the acetabular components by doing finite element analysis upon different orientation data obtained from the first stage. The last stage is proposing the new composite material with the aim to study their properties at different weightage percentages and to compare their results based on FEA study.

4.2 Numerical and Mathematical Modeling for Acetabular Cup Orientation

Fumihiro et al. [1] have developed a new mathematical formula on defining a suitable safe zone for acetabular cup orientation with five parameters applied (refer to Sect. 2.3.1). These parameters are considered as a pivot and must be complied respectively to get the result of acetabular cup orientation that will comply with the safe zone orientation. Most of the studies show that the cup orientation of any acetabular cup is discovered by simulation software [2, 3] and some others from experimental results [4]. Thus, this research aims to combine both methods' results and simultaneously run the mathematical formula [1] in MATLAB with additional judgment from Widmer et al. [2, 5] and Klingenstein et al. [6] on defining the safe zone.

Run in by using MATLAB, a function is created with the input arguments vary for theta (θ), head-neck ratio, cup inclination (α) and cup anteversion (β). All input parameters will be commanded with multiple values of cup anteversion ranges; hence

M. F. b. A. Manap et al., *Total Hip Replacement (THR)*, SpringerBriefs in Applied Sciences and Technology, https://doi.org/10.1007/978-981-96-0975-8_4

```
function | theta,E,D,FL,EXT,E2,D2,ER,IR,E3,D3,ABD,ADD] = TotalLatest(AngleMax,✔
necktodiameter,alpha,beta)
global a b
a = 52; %fixed at a rate of 52 degree
b = 20; %fixed at a rate of 20 degree

%input argument:
%AngleMax = Total angle of acetabular cup
%Neck = Neck width at impingement (diameter of ball)
%Diameter = Diameter of the head (diameter of neck)
%Alpha = Acetabular cup inclination
%Beta = Acetabular cup anteversion

%output argument:
% theta = Oscillation angle
% FL = Flexion angle
% EXT = Extension angle
% ER = External Rotation angle
% IR = Internal Rotation angle
% ABD = Abduction angle
% ADD = Adduction angle
```

Fig. 4.1 The input and output argument function created upon Fumihiro equation in the MATLAB command space. The values 'a' and 'b' were set as global input [1]

resulting in many outputs comprised of flexion, extension, external rotation, internal rotation, abduction and adduction. Referring to Fig. 4.1, it shows the function created in the MATLAB command space. The value of cup anteversion ranged from 2° to 50° with two-degree increments. Five conditions were applied with oscillation angle (OsA) values of 110°, 130°, 135°, 140° and 145°, respectively. Based on Eq. 4.1, the head-neck ratio plays a vital role; thus a femoral head of 22 mm diameter with a neck width of 12 mm is chosen as the first criteria for 120 degrees OsA with an output ratio of 1.83. The other conditions are by combining 28, 32, 36 and 40 mm femoral head diameter with a constant 12 mm neck width. These will yield a head-neck ratio of 2.33, 2.67, 3.0 and 3.33, respectively.

Based on previous literature review, new parameters needed for the range of motion were discovered. For all the criteria, it is intended to get optimum values of Flexion 110°, Extension 30°, External Rotation of 40°, Internal Rotation 120°, Abduction 30° and Adduction 40°. These new parameters are set based on the previous work conducted by other researchers which is the range of motion which are too subjective to be defined as general [5, 6]. The new parameters of ROM used for the boundary condition were referred from the results by simulation [2] and computerized range of motion analysis [6]. Some research stated that internal rotation at 90° flexion is meant by adding the values of flexion to the intended range of internal rotation [3] as simulated visually. However, based on the previous work done [6], 120° of internal rotation is the minimum allowable range of motion that satisfies the ADL of the patient; thus the parameter is chosen for the numerical study. From the parameters set at the optimum range of motion, the nearest values of output angle (flexion, extension, internal rotation, external rotation, abduction, adduction) shown in MATLAB are chosen and transferred to the Excel Workbook. This acts purposely to ease the process of determining the inclination angle (α) and the anteversion angle (β) that will match the parameters set before. These boundary conditions are carefully plotted in graph order to get the common ranges that complies with all parameters stated.

The ranges of radiographically cup inclination angle are chosen from 0° to 60° degrees with 5° degrees intervals that enable the calculations of all the output range of motion mentioned before. The cup anteversion angles chosen from range 0° to 50° with 2° interval are due to minimizing the error as the aim is to find the nearest value of range of motion from MATLAB. In other words, for every 5° of inclination angle (α), there should be various anteversion angles (β) ranging from 0° to 50°. The output from the equations that were run in MATLAB will be in terms of various angles of ROM. However, only certain anteversion angle (β) that obeyed the intended ROM had been chosen for that criteria. Although the calculation seems complicated, MATLAB software enables the calculation of every input angle without taking too much time by commanding to replace the anteversion angle (β). Every formulation mentioned before (Eq. 2.1–2.13) was run separately in MATLAB in order to get the value of inclination angle (α) and anteversion angle (β) that are nearest to our targeted ROM. As a result, the variation of the six basic range of motions being set as the new parameters can be concluded.

Here, an example of calculating the FL110 is explained as follows. In order to draw the graph of the FL110 that intends a minimum of 110°, the estimation value method consideration was executed. Table. 3.1 shows an example of a definition of an appropriate inclination angle and anteversion angle using the 2.67 head-neck ratio. 2.67 head-neck ratio is the ratio of femoral head diameter and femoral neck diameter with 32 mm femoral head and 12 mm femoral neck, respectively. Take an example of 10° inclination angle with a various range of anteversion angle from 30° to 2° with two interval exhibits that the minimum value of FL100 falls into the area of 26° anteversion angle. Thus, the value of 10° inclination angle antevert with 26° anteversion angle is chosen as one of the points from the graph that will be plotted (highlighted in yellow).

The ranges showed are from 30° to 2° anteversion only although the real calculation in MATLAB is from the range of 50° to 2°. Table. 4.1 shows a simplification for the reader viewing our estimation value of the graph. The N/A symbol shown in the table represents that no value appeared upon the formula that has been used in MATLAB and the values reported are rounded-off into three significant figures.

According to [7], the values of 'a' and 'b' are technically varied depending on the design of the femoral stem. In this case, it sticks on previous studies' statement that it must be determined as global input with values of 52° degrees and 20° degrees, respectively. It is to reduce the complexity of the function developed in MATLAB® which means that stem-neck angle (CCD) is fixed. This control condition is compulsory to ensure the output parameters determined are not affected by the different values of 'a' and 'b' such as the desire to get different output of parameters that are set earlier. It is beyond the scope of this study if the arbitrary constants of 'a' and 'b' are varied with numerous combinations of the study. The definition of the values

Table 4.1 The screenshot of FL110 estimation angles is shaded with yellow color. The first value that reached 110 is selected as the minimum value allowable for the safe zone orientation. (This is the example FL110 for 2.67 head-neck ratio)

A (degree)	B (degree)														
	30	28	26	24	22	20	18	16	14	12	10	8	6	4	2
5	111.085	109.140	107.190	105.236	103.278	101.315	99.348	97.377	95.401	93.421	91.437	89.449	87.456	85.459	83.458
10	114.591	112.676	110.753	108.821	106.880	104.931	102.974	101.009	99.037	97.056	95.068	93.072	91.068	89.057	87.038
15	118.210	116.305	114.386	112.454	110.509	108.552	106.584	104.605	102.614	100.613	98.601	96.578	94.545	92.502	90.448
20	122.015	120.099	118.163	116.210	114.239	112.253	110.252	108.237	106.207	104.165	102.109	100.041	97.959	95.864	93.756
25	126.093	124.143	122.169	120.171	118.152	116.113	114.056	111.982	109.892	107.786	105.665	103.528	101.377	99.210	97.028
30	130.553	128.546	126.507	124.439	122.345	120.227	118.088	115.929	113.752	111.557	109.345	107.116	104.871	102.608	100.327
35	135.551	133.454	131.318	129.146	126.944	124.715	122.461	120.185	117.889	115.574	113.242	110.891	108.523	106.135	103.727
40	141.325	139.089	136.807	134.485	132.128	129.742	127.329	124.893	122.438	119.964	117.472	114.963	112.435	109.886	107.314
45	148.313	145.839	143.321	140.764	138.175	135.560	132.921	130.264	127.590	124.901	122.199	119.481	116.745	113.989	109.207
50	157.587	154.567	151.568	148.578	145.591	142.604	139.617	136.630	133.642	130.654	127.663	124.666	121.657	118.630	115.575
55	N/A	N/A	N/A	159.590	155.604	151.825	148.175	144.615	141.119	137.672	134.260	130.870	127.488	124.099	120.683
60	N/A	N/A	N/A	N/A	N/A	167.172	160.945	155.818	151.198	146.871	142.735	138.726	134.795	130.898	126.990

Fig. 4.2 Represent the left THR. 'a' is the angle of neck position at horizontal plane and 'b' is the anteversion of neck around vertical axis from coronal plane (reproduced from [1], with permission)

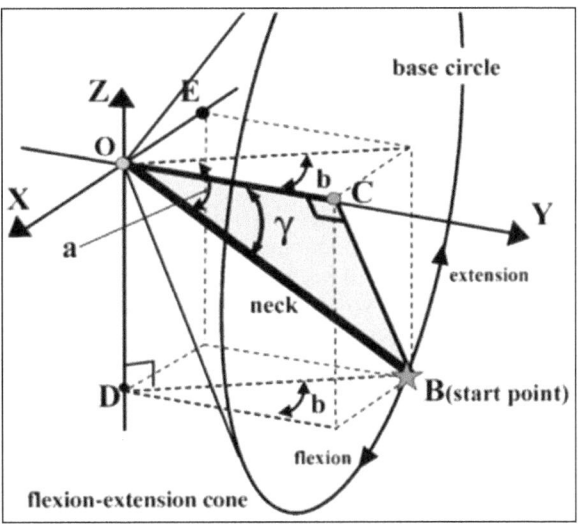

'a' and 'b' in the function developed as global value in MATLAB® can be defined as in Fig. 4.2.

4.3 Computer Simulation of Acetabular Components

The bearing connection of the femoral head and cup is usually modeled with a ball-in-socket geometry and is highly conforming. Radial clearance is value of $c = R_2 - R_1$ in which the R_1 and R_2 represented the head and cup radii, respectively. Due to complication of ball-in-socket geometry, some papers used a simple ball-on-plane configuration adopted from Hertzian Contact Theory simply defined by the effective radius R' and elastic modulus E' as in Eqs. 4.1 and 4.2 [8].

$$\frac{1}{R'} = \frac{1}{R_1} - \frac{1}{R_2} = \frac{c}{R_1(R_1 + c)} \tag{4.1}$$

$$\frac{1}{E'} = \left(\frac{1 - v_1^2}{E_1} + \frac{1 - v_2^2}{E_2} \right) \tag{4.2}$$

From these two equations, the value of deformation, δ could be found based on Eq. 4.3 in which 'F' is the force exerted in Y-axis direction.

$$F = \frac{4}{3} E' R'^{1/2} \delta^{3/2} \tag{4.3}$$

The Hertzian Contact Theory is an important validation factor in doing the finite element as the total deformation results in FEA will be compared at $\alpha = 0°$. The other geometrical parameters such as the inclination and anteversion are not included in this simplified model calculations.

Finite element analysis (FEA) will be run in ANSYS WORKBENCH as the software enables to run complexity project. First, a design of the acetabular component implant will be drawn on the SolidWorks 3D Drawing, which included the femoral head, acetabular cup and metal backing. The dimension of these particular acetabular components will be determined based on the data per agreed from the numerical approach results. However, for the purpose of defining the implant in simulation, three combinations of acetabular components that are mostly used in orthopedic surgery are used for the FEA. All other parameters are constant with respect to the femoral head size. The 28, 32 and 36 mm femoral head diameter sizes were selected only for the analysis. For example, single out femoral head size of 28 mm diameter, the thickness of the acetabular cup will be 7 mm and metal backing will make about 4 mm with respect to the femoral head size. These will make the outer diameter of acetabular cup of 42 mm and outer diameter of metal backing with 50 mm. On the other hand, taking femoral head size of 36 mm, the thickness of acetabular cup and metal backing are maintained identical with the sizes of 7 mm and 4 mm, respectively. The only difference is the outer diameter size of the acetabular cup that will make 50 mm and metal backing outer diameter size of 58 mm. In the case of using 32 mm femoral head, it also has the parametric similar to the 28 mm femoral head and 36 mm femoral head size. Refer to the schematic drawing of the three types of acetabular components in Fig. 4.3.

The first step is by drawing all the acetabular components in SolidWorks with three components. Figure 4.4 is the screenshot of the femoral head being drawn as a solid ball in SolidWorks.

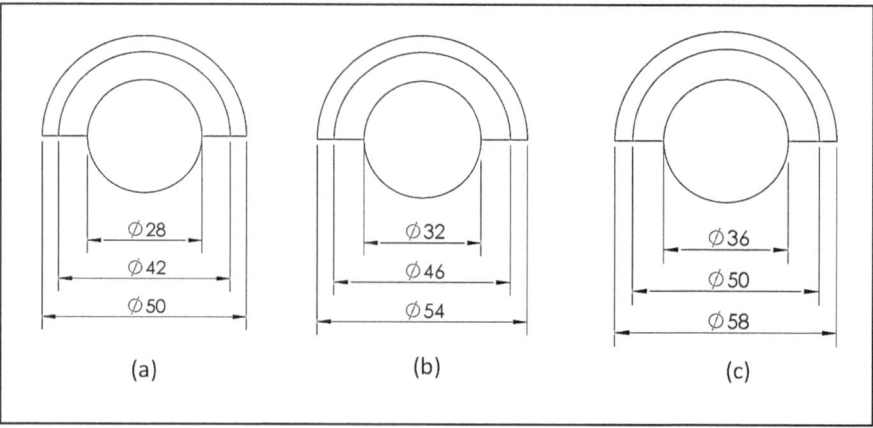

Fig. 4.3 The schematic drawing of three types of acetabular components. **a** 28 mm femoral head diameter; **b** 32 mm femoral head diameter; and **c** 36 mm femoral head diameter (author's illustration)

Fig. 4.4 Example of the head with 28 mm diameter (scale 1:1)

The femoral head being used in this simulation analysis is not included with the groove hole for the femoral stem placement. It is to reduce the complexity of the simulation approach as the forces are given to the head with respect to the center of the head only. It is also important that the analysis aimed to achieve the acetabular components only. For both combinations, the thickness is identical as previously mentioned with only the nominal size being different with respect to the femoral head size. The diagram in Fig. 4.5 shows the trimetric view of the acetabular cup liner with 180° hemisphere shape.

Although there are many designs and types of acetabular cup depending on the supply of company preferences, a general condition for this study is selected which means the critical parameters, the thickness and the half-hemisphere shape are considered. For the metal backing, we also make assumptions about the general design as in Fig. 4.6.

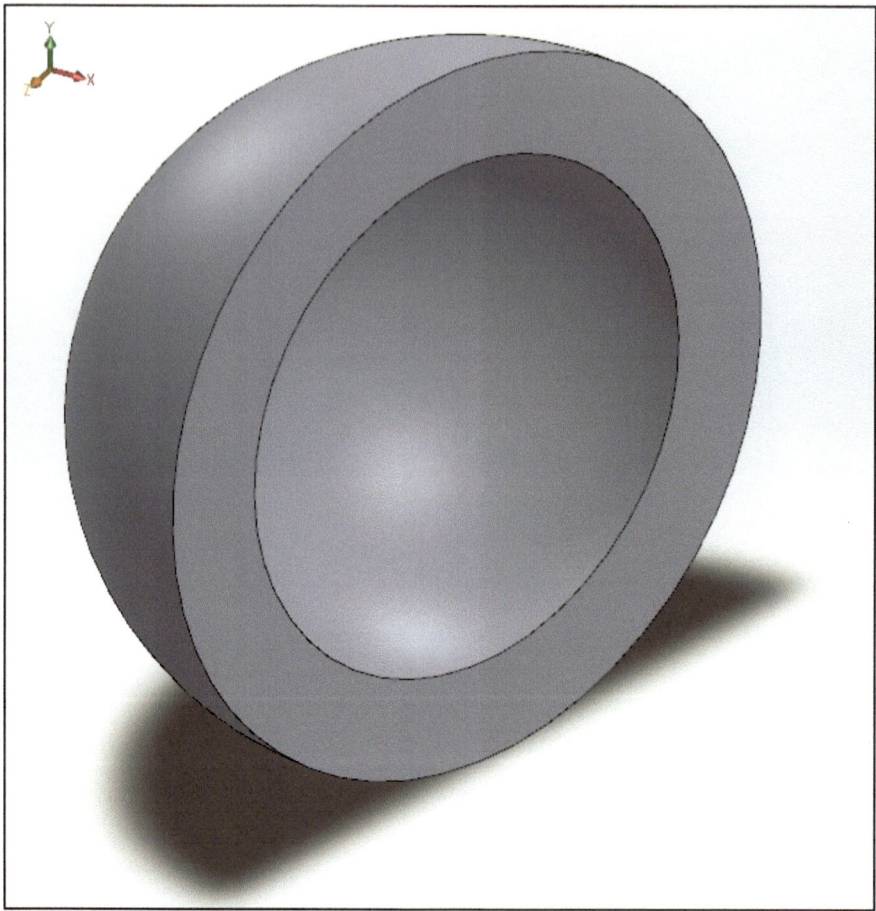

Fig. 4.5 Example of acetabular cup (liner) with thickness of 7 mm (scale 1:1)

After drawing the three components separately, they are mated together with reference to various inclination angles (α) and anteversion angles (β). All were defined differently based on analytical results at different acetabular component sizes. Depending on various previous studies [9–11], a setting with inclination of 0°, 50°, 60° and 70° as the basic orientation that is needed when dealing with simulation analysis is chosen for the comparison with Korduba et al. [10]. In the case of the combination of orientation that includes the anteversion angle (β), the angle for the inclination angle and anteversion angle derived depends on the safe zone diagram from the analytical results. The important thing is that the orientation of the respective acetabular components needs to be set first during the drawing phase in SolidWorks before converting it into IGS files. SolidWorks is a powerful software that has special command namely TRIAD command as shown in Fig. 4.7 that will allow anteverting the acetabular cup with any angle that is compulsory for this study.

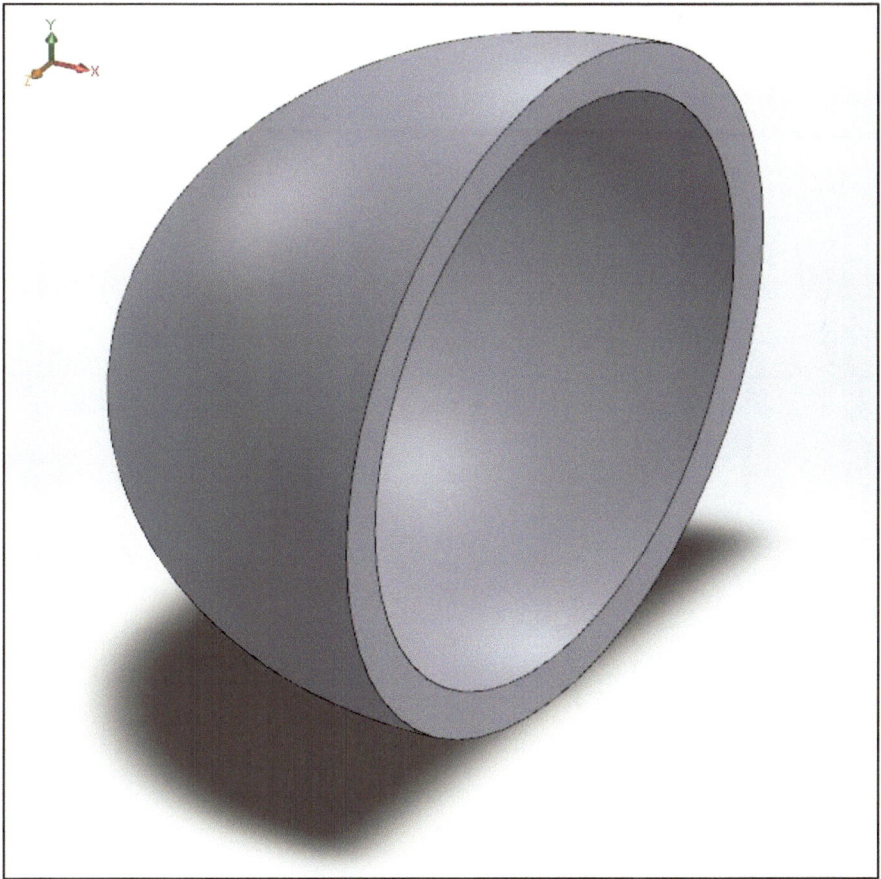

Fig. 4.6 Metal backing that is being used as the cover for the acetabular component with 4 mm thickness (scale 1:1)

It is compulsory to ensure that the TRIAD command applied is based on the origin of the acetabular cup, not the whole set of the acetabular component.

Typically hard-on-soft bearing (MoP) connections use the 28 mm femoral head diameter with various acetabular cup and metal backing sizes. In this simulation, the 28 mm femoral head diameter, 7 mm thickness acetabular cup and 4 mm metal backing with overall sizing of 50 mm was drawn into the SolidWorks.

Figure 4.8 illustrates the example of orientating the angle for the 28 mm femoral head diameter with imaginary angle line (red color). Then, all the parts are assembled with respect to the orientation intended for this study.

Technically, there are 15 models of acetabular components with different orientations and sizes being analyzed in ANSYS WORKBENCH and Fig. 4.9 shows an example of 28 mm femoral head case. The 'Maximum Safe Inclination Angle' of all three types of acetabular components was retrieved from the numerical results data.

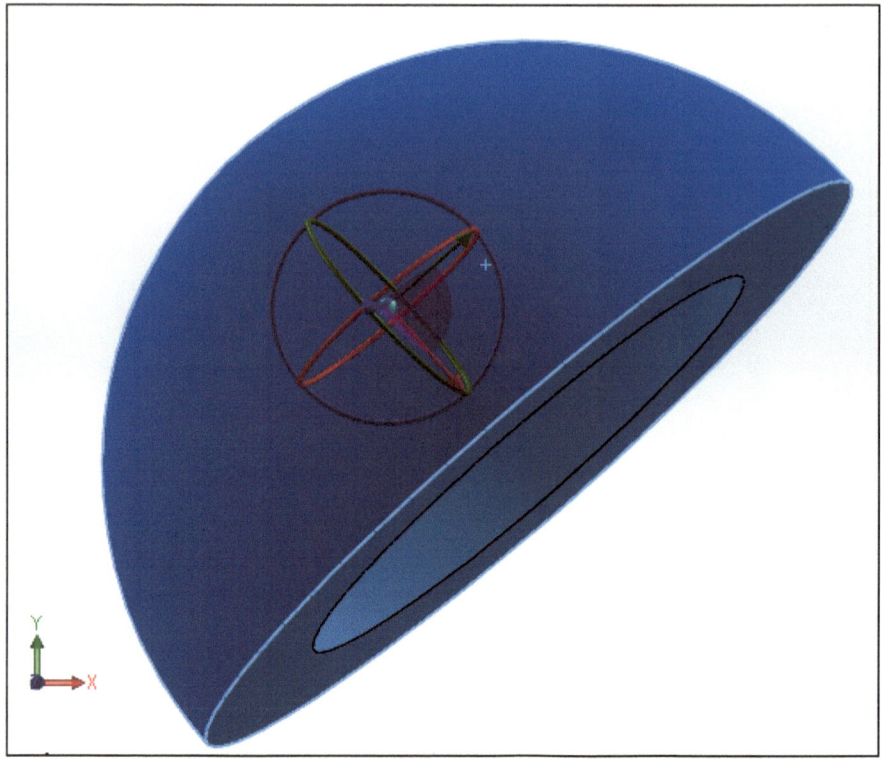

Fig. 4.7 Using TRIAD command to set the anteversion angle of acetabular cup

After all the acetabular components are assembled with the orientation needed for the analysis using FEA, the files from SolidWorks are imported to ANSYS WORK-BENCH. It is important to ensure the files are formatted in IGS format only upon transferring to ANSYS software as other formats are not compatible. After being converted into IGS and imported to ANSYS software, STATIC STRUCTURAL is the project schematic that will be used upon analyzing the performances of acetabular components.

4.3.1 Engineering Data and Meshing

The ANSYS WORKBENCH software is equipped with engineering data for the common materials in this world; however, the UHMWPE is not precisely shown. Only polyethylene (PE) data is provided in the database and the material properties are different with UHMWPE. Thus, this study will emulate the data from [12] in order to run the simulation with only two mechanical properties being sufficient which are

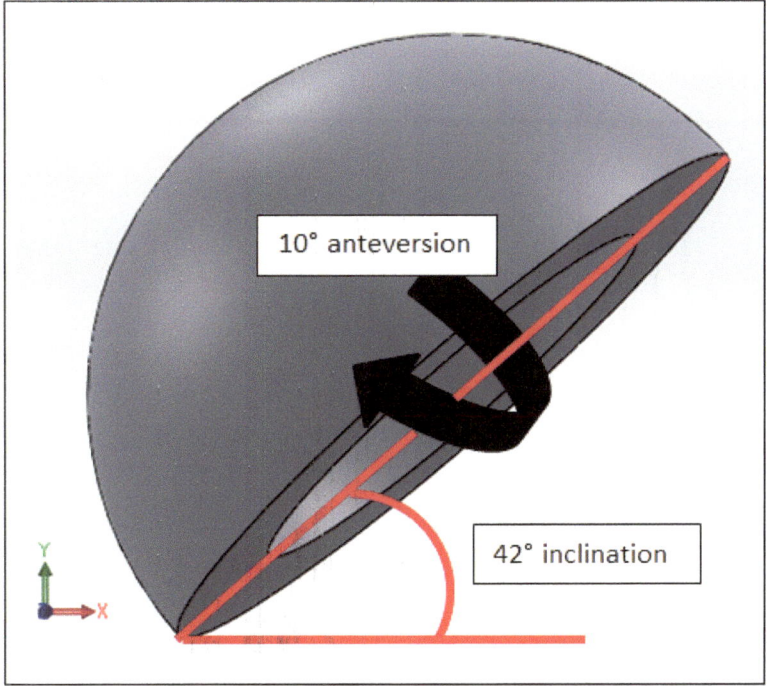

Fig. 4.8 The example of 3D model of UHMWPE acetabular cup upon using 28 mm femoral head size with 42° inclination angle and 10° anteversion angle

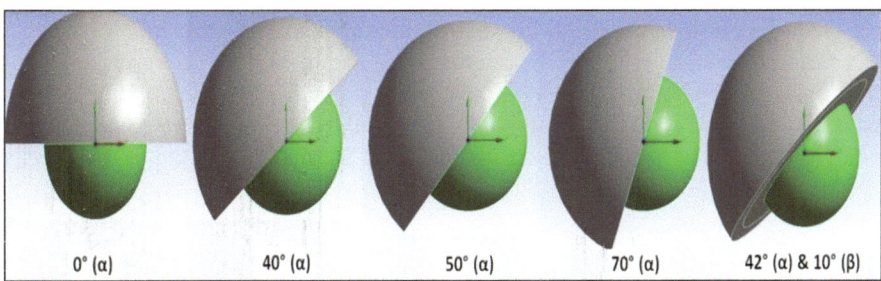

Fig. 4.9 The example of 28 mm femoral head diameter cases with 5 different orientations for analysis studies using FEA

the Young Modulus and Poisson's Ratio. As mentioned previously, the other two parts are considered rigid as these two parts have higher stiffness; thus no deformation may happen on the femoral head and metal backing [10, 13]. By assuming the other two components as rigid, the computational time elapses for the results in FEA can be reduced. The acetabular cup materials properties are dictated based on previous study that uses the Young Modulus and Poisson's Ratio with 945 MPa and 0.45, respectively

Table 4.2 The meshing consisting of element and nodes is shown upon 15 models of acetabular components using FEA

Femoral head diameter (mm)	Orientation(°)	Number of elements	Number of nodes
28	0	12,364	25,499
	40	12,290	25,465
	50	11,873	24,755
	70	13,462	27,152
	Maximum safe inclination angle	11,921	24,837
32	0	15,638	32,343
	40	15,543	32,221
	50	15,558	32,226
	70	15,551	32,282
	Maximum safe inclination angle	15,642	32,324
36	0	19,233	40,056
	40	20,573	41,654
	50	19,167	39,888
	70	19,462	40,273
	Maximum safe inclination angle	19,226	39,958

[12]. The meshing is set with 10-noded solid tetrahedral elements, and the size of the elements between the contact region of femoral head and acetabular cup was set at 1 mm. The contact between the femoral head and the acetabular cup is defined as frictionless as this study focuses on the contact pressure upon the articulate surface between these two components. Table. 4.2 shows the meshing statistic recorded upon doing 15 models of acetabular components in FEA.

4.3.2 Boundary Conditions and Loading Conditions

The setup boundary condition for all cases was executed through the ANSYS WORK-BENCH software. In this study, the loading condition is given with 2450N forces upon center of femoral head vertically. The loading condition is based on the reference from Korduba et al. [10] that represents about two or three times of typical human body weight. Different gaits give different loading conditions, yet this research shows the average force applied upon normal human body in which the force exerted is sufficient and relevant to this parametric study. Meanwhile, fixed support is held on the outer surface of acetabular cup shell that will constrain the movement of the cup in

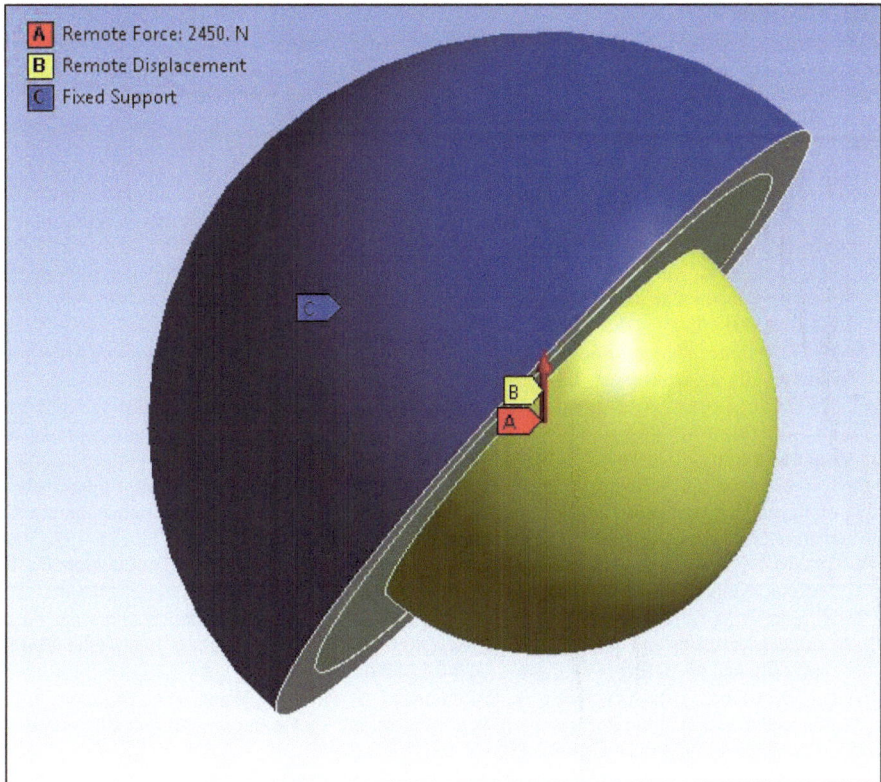

Fig. 4.10 The loading condition and constrained parts of the model

any direction as in Fig. 4.10. Remote displacement is also applied to the femoral head that will halt any possible rotation movement when the two articulate surfaces meet with each other.

From the data recorded upon this analysis, the values of Von-Mises stress, Total Displacement and Contact Pressure value will be investigated and analyzed. After that, the results will be set as a controlled value. Then, from the experimental methodology, new value of Young modulus and Poisson's Ratio of the new composite material samples will be figured out. By using the experimental value results of the new composite materials, the FEA will be re-run with the same model orientation as from the controlled parameters; the analysis will be recorded and compared with the intention to investigate whether the new composite material may improve the implant in terms of three mechanical analysis.

References

1. F. Yoshimine, K. Ginbayashi, A mathematical formula to calculate the theoretical range of motion for total hip replacement. J. Biomech. **35**(7), 989–993 (2002)
2. K.-H. Widmer, M. Majewski, The impact of the CCD-angle on range of motion and cup positioning in total hip arthroplasty. Clin. Biomech. (Bristol, Avon) **20**(7), 723–728 (2005)
3. W.-T. Ji, K. Tao, C.-T. Wang, A three-dimensional parameterized and visually kinematic simulation module for the theoretical range of motion of total hip arthroplasty. Clin. Biomech. (Bristol, Avon) **25**(5), 427–432 (2010)
4. N.J. Little, C.A. Busch, J.A. Gallagher, C.H. Rorabeck, R.B. Bourne, Acetabular polyethylene wear and acetabular inclination and femoral offset. Clin. Orthop. Relat. Res. **467**(11), 2895–2900 (2009)
5. K.-H. Widmer, A simplified method to determine acetabular cup anteversion from plain radiographs. J. Arthroplast. **19**(3), 387–390 (2004)
6. G.G. Klingenstein, A.M. Yeager, J.D. Lipman, G.H. Westrich, Computerized range of motion analysis following dual mobility total hip arthroplasty, traditional total hip arthroplasty, and hip resurfacing. J. Arthroplast. **28**(7), 1173–1176 (2013)
7. F. Yoshimine, The influence of the oscillation angle and the neck anteversion of the prosthesis on the cup safe-zone that fulfills the criteria for range of motion in total hip replacements. The required oscillation angle for an acceptable cup safe-zone. J. Biomech. **38**(1), 125–132 (2005)
8. T. Sato, N. Sato, Clinical relevance of the hip joint: Part II–Importance of joint distraction. Int. Musculoskelet. Med. **37**(4), 141–145 (2016)
9. X. Hua, B.M. Wroblewski, Z. Jin, L. Wang, The effect of cup inclination and wear on the contact mechanics and cement fixation for ultra high molecular weight polyethylene total hip replacements. Med. Eng. Phys. **34**(3), 318–325 (2012)
10. L.A. Korduba, A. Essner, R. Pivec, P. Lancin, M.A. Mont, A. Wang, R.E. Delanois, Effect of acetabular cup abduction angle on wear of ultrahigh-molecular-weight polyethylene in hip simulator testing. Am. J. Orthop. **43**(10), 466–471 (2014)
11. R.K. Korhonen, A. Koistinen, Y.T. Konttinen, S.S. Santavirta, R. Lappalainen, The effect of geometry and abduction angle on the stresses in cemented UHMWPE acetabular cups--finite element simulations and experimental tests. Biomed. Eng. Online **4**(1), 32 (2005)
12. D. Kluess, H. Martin, W. Mittelmeier, K.-P. Schmitz, R. Bader, Influence of femoral head size on impingement, dislocation and stress distribution in total hip replacement. Med. Eng. Phys. **29**(4), 465–471 (2007)
13. H. Jiang, Static and dynamic mechanics analysis on artificial hip joints with different interface designs by the finite element method. J. Bionic Eng. **4**(2), 123–131 (2007)

Chapter 5
Experimental Work of Epoxy-UHMWPE Materials

Abstract In this study, it is intended to incorporate the UHMWPE into the polymer matrix of Epoxy (Ep). These new composite materials are suggested to perform better than UHMWPE in terms of mechanical properties. Yet, in the study of total hip replacement, although UHMWPE proved to be a successful material selection for the acetabular cup, it still needs a lot of improvement to improve the implant life.

5.1 Experimental Work of Epoxy-UHMWPE Materials

In this study, it is intended to incorporate the UHMWPE into the polymer matrix of Epoxy (Ep). These new composite materials are suggested to perform better than UHMWPE in terms of mechanical properties. Yet, in the study of total hip replacement, although UHMWPE proved to be a successful material selection for the acetabular cup, it still needs a lot of improvement to improve the implant life.

The material is to be fabricated as the liner material must be able to withstand high compression load, edge-loading, resist deformation and minimal friction against the femoral head. Before heading to the sample preparation, the coding of the formulation will be tabulated for easiness on reading the composition value. The composite intended to be used as the substitute material is Epoxy/UHMWPE. The formulations of the polymer composite materials are shown in Table 5.3. Epoxy resins were obtained from the Faculty of Applied Science, UiTM with the coding of Morcote BJC-29 supplied by Vistec Technology Sdn Bhd. Meanwhile, UHMWPE graded GUR 4120 was supplied from Ticona Engineering Polymer, China in powdered form with molecular weight of 5×10^6 gmol^{-1} (Table 5.1).

Here, the sample example on the calculation of the coding is as follows (Equation 5.1–5.4):

EpUHMWPE5 (95% resin and 5%UHMWPE) will be the benchmark for calculating the weight of the Epoxy resin and the required amount of UHMWPE. Density meter will be used in the measurement of Epoxy, ρ_s and the obtained value is 1.21 gcm^{-3}. To calculate the respective amount of Epoxy and UHMWPE, estimation method based on mold volume will be carried out. The mold volume V_m was

M. F. b. A. Manap et al., *Total Hip Replacement (THR)*,
SpringerBriefs in Applied Sciences and Technology,
https://doi.org/10.1007/978-981-96-0975-8_5

Table 5.1 Sample
formulation coding of the
variation of filler loading in
polymer Epoxy matrix with
UHMWPE

Coding	Matrix (wt %)	Fillers (wt %)
		UHMWPE
EpUHMWPE0	100	0
EpUHMWPE1	99	1
EpUHMWPE3	97	3
EpUHMWPE5	95	5
EpUHMWPE7	93	7
EpUHMWPE10	90	10

measured by the dimension of the cavity which is the value of length x width x
height. For this experiment to be carried out, the mold volume of (80 x 10 x 10) mm
is 80 cm^3. Mass will be determined from the density equation denoted as;

$$\rho = \frac{m}{V} \tag{5.1}$$

Here, the m is the mass in g and V is the volume in cm^3, thus the estimation mass of
Epoxy to be used in this experiment can be written as;

$$\rho_s = \frac{m_s}{V_s} \tag{5.2}$$

where m_s = mass of the Epoxy and substituting the value will yield;

$$1.21 = m_s/80 \tag{5.3}$$

Thus, the mass of the Epoxy will be used in this coding combination is,

$$m_s = 96.8\text{g} \tag{5.4}$$

This yield value is the theoretical value of Epoxy mass and a recovery factor
of about 5% must be considered as the residue on the disposal cup container used.
Thus, the final weight of the Epoxy required = 101.64 g. Note that the recovery factor
must be included in every coding of the samples that will be run in this experimental
approach. The weight of the UHMWPE will be measured by the reference value of its
own coding Epoxy and in this case, the value is 101.64 g and will be multiplied by 5%
that will yield about 5.082 g of UHMWPE. However, there is another consideration
as the Epoxy used is in two parts (Morcote BJC-29) and from the data sheet, the
mixing ratio is 3.8:1.8 for Epoxy and hardener, respectively. Thus, the calculation of
the ratio will show that the amount of Epoxy = 65.521 g and hardener = 31.0365 g.

For the UHMWPE as the benchmark materials, the output results from the simulation approach were set as controlled parameter. This new composite system variation is suggested to improve the implant life, especially on the acetabular component

region. In order to apply these materials selection to the FEA, only two main values are required which are the Young Modulus and Poisson's ratio of the materials. The Young Modulus of all the specimens will be determined as the input engineering data for the simulation using FEA in which the values are calculated based on mechanical testing that will be performed later.

5.1.1 The Gel Time of Epoxy

Gel time is defined as the total time for the Epoxy and hardener to form a gel under constant stirring condition. This process is measured visually as the physical evidence shows the mixture turns into a gel. Stop watch will record the time taken for the mixture to change state into a gel. This procedure can be illustrated in Figure 5.1. As a standard procedure, the test will be carried out in room temperature. To find the gel time, only 50 g of total Epoxy which is made out of 33.929 g Epoxy and 16.071 g hardener was used.

From time to time, the mixture will be raised upon the cup in order to determine the viscosity of the resin. The flow of the mixture should be continuous with respect to gravity and the time taken will be recorded. If the mixture is not continuous, the stirring process needs to be done again.

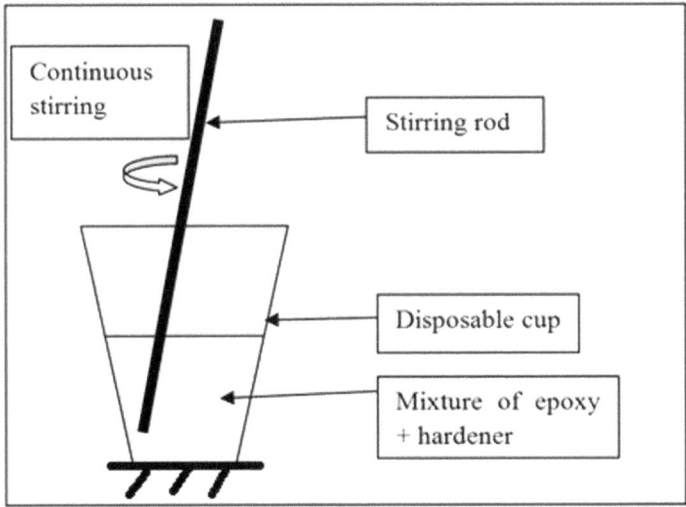

Fig. 5.1 Set-up the technique in identifying the gel time of the Epoxy resins

5.1.2   Pressure-Assisted Hand-Layup Techniques

The mold of 80 cm^3 that will be used for pouring the mixture was made from '3 mm vinyl plastic' which was prepared using cutter and ruler. The main guide of this process is the gel time requires about 35 min for the Epoxy and hardener to mix homogeneously. In order to complete the mixing of Epoxy, hardener and UHMWPE in a proper manner, it is important to divide the time of every step required from the gel time results. Figure 5.2 shows the apparatus that was used to prepare the sample.

From the gel time benchmark, the mixing of the UHMWPE and the Epoxy matrix was carried out by squashing the mixture using hand. First, the Epoxy and hardener are mixed together and constant stirring methods are utiliZed for about 20 min. This is the possible method given that the restricted time frame as the gel time of the Epoxy is around 35 min. Then, instantly mix UHMWPE with Epoxy/hardener in which it continues stirring for 10 min with vigorous stirring as to allow the UHMWPE to mix homogeneously into Epoxy matrix. Given that the gel time is about 35 min, so roughly 5 min is given for the mixture to cure properly.

A hydraulic press (Figure 5.3) will be used to compress the composite materials as desired to get the samples with a dimension of (80 × 100 × 10) mm. The top and the bottom platen will be compressed for about 40 min upon this hydraulic press machine. This method will ensure that every sample produced is free from clear view warped due to mold condition and heat exposed to the materials during curing process.

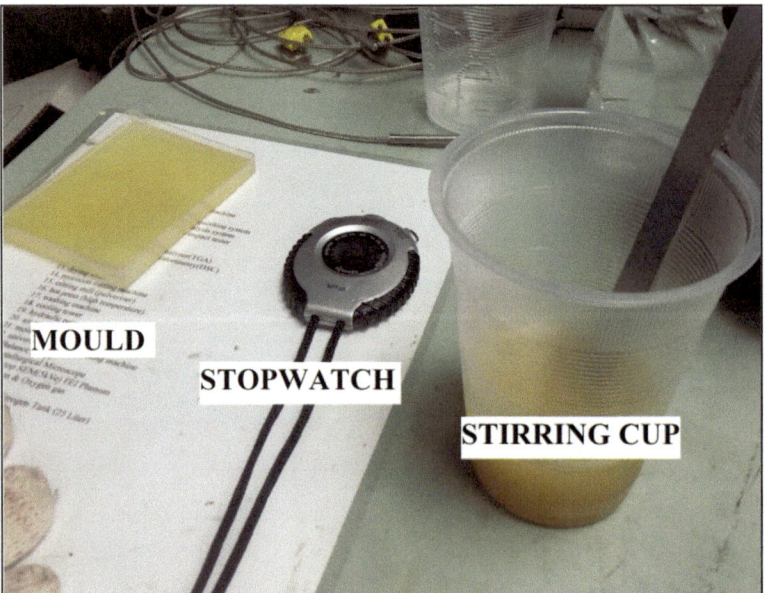

Fig. 5.2   The basic apparatus and the mold being used for the sample preparation

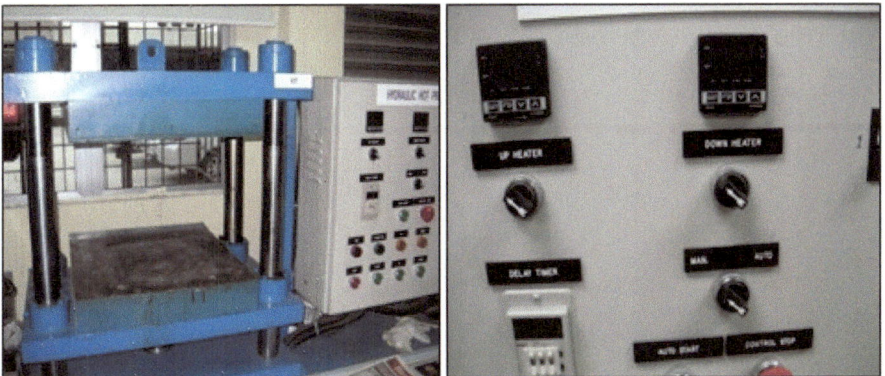

Fig. 5.3 The hydraulic press machines purposely in assisting the sample fabrication

Note that in automatic mode, the bottom platen will displace downward as per set. These techniques particularly ensure that every sample is removed from the compression load at the same time to avoid any variation on the samples.

5.1.3 Conditioning of the Prepared Samples of EpUHMWPE

As the complete form sample was obtained from the fabrication process, the sample was then subjected to post-curing process. The post-curing process required the sample to be heated to a specific temperature with a specific time, and for this case, the sample was put inside an air-circulated oven at 70 °C for 24 hours. By applying the post-curing process, the curing process inside the sample can be considered to fully take part to form the desired composite sample. There are some precautions needed during the post-curing process that is the sample inside the oven needs to be loaded on the top with a flat piece of metal (the dimension must cover the whole surface area of the sample) to ensure warping will not occur during the post-curing process. After finishing the post-cure process, the sample was taken out from the oven and was left to cool to room temperature before it was cut to the required dimension as required for the testing and analysis later on.

The precision cutting machine is fabricated and tailored to cut the thermoset composite due to the design of the rotational blade (diamond-embedded tile-cutting blade). This blade does not have any sharp edges, and thus the cutting mechanism is suspected to be more like grinding and shearing. This mechanism of cutting was found to be the best as there is no evidence of chipping off from the cut sample. Besides, there are no cutting edges observed from all the samples yielding to elimination of weak points on the side of the sample. All the cut samples were then properly labeled and kept inside a sealed plastic bag as to wait for the testing to be carried out. The

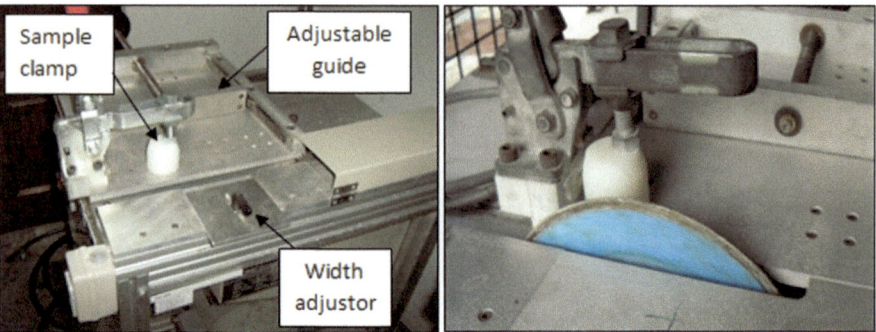

Fig. 5.4 The fabricated precision cutting machine used to cut the sample to the required width and length for the tests being carried out

precision cutting machine has three mechanisms that will ensure the cutting of the composite samples is at the pre-set dimension all the time as shown in Figure 5.4.

The said mechanisms are the sample clamp, the adjustable guide and the width adjustor. This mechanism will work hand in hand to ensure the sample to be cut will stay in place and will be cut to the accordingly setting value of the width adjustor.

The completed samples will undergo the optimization test of compression as the aim of the sample is to define the Young modulus of the new composite material. In terms of Poisson's Ratio, value from the engineering data of Epoxy was selected at every variation as Epoxy is claimed as the dominant material; thus will not result in a significant difference. Previous studies were conducted and it was found that in the case of using UHMWPE for simulation studies, due to the small magnitude, it is often to assume the value of Poisson's Ratio of about 0.4 [1]. The results from each sample will be commanded as an input of new materials into the engineering data of ANSYS WORKBENCH software. Then, static structural analysis will be run by the new data information and the results from the FE will be analyzed and recorded.

5.1.4 Compression Test for EpUHMWPE Young's Modulus

The compression test was done to find the Young's Modulus value. The Young's Modulus is one of the main criteria needed in the input engineering data when re-running the ANSYS WORKBENCH. There are at least compulsory two mechanical properties when dealing with the material properties in ANSYS WORKBENCH which include the Poisson's Ratio. For this compression test, each variant is prepared with three test specimens. The specimens had a parallel-piped form with a square section according to ASTM D695-96. After each variant samples of Ep/UHMWPE was done according to the methodology, the samples were cut with the dimension as in Figure 5.5.

Fig. 5.5 Dimension of the
compression test sample
(unit in mm)

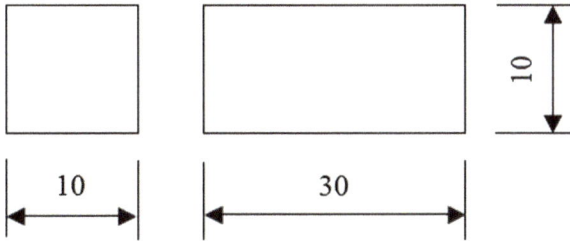

Figure 5.6 shows one of the variants that have been cut to run the compression test. The specimens are the EpUHMWPE3 with the amount of 3% UHMWPE and 97% of Epoxy based on weightage measurement ratio.

The samples were loaded into a universal testing machine of Shimadzu brand which is possessed by Faculty of Applied Science, Universiti Teknologi MARA with the machine code of UiTM/PS01/A/090108/20080001118. The bottom plate was fixed and the top plate is mobile as in Figure 5.7. Test was conducted under ambient laboratory conditions, 23 ± 1 °C. The load was applied at a constant displacement rate of 1.3mm/min with the stroke of 10% height of the sample. TrapeziumX is the built-in software that was included while running the testing machine. Due to

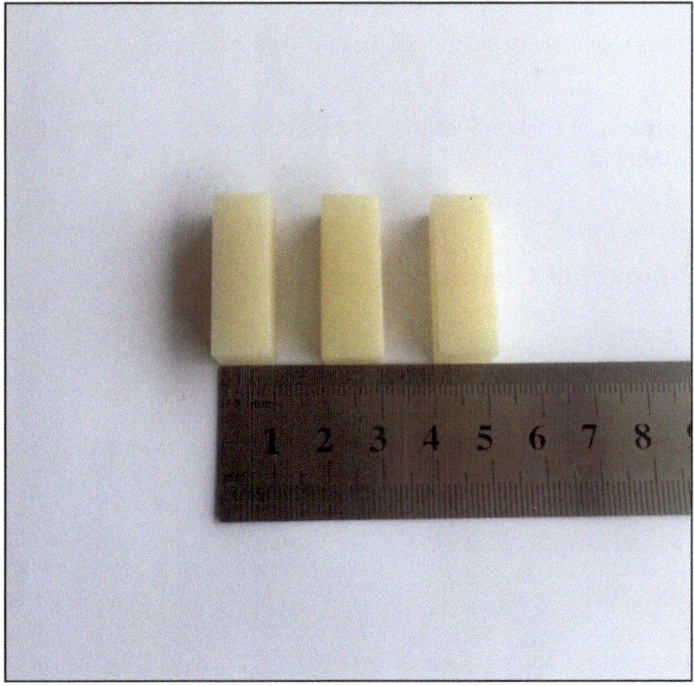

Fig. 5.6 EpUHMWPE3 sample for the compression test (unit in mm)

Fig. 5.7   Compression Testing Device used for EpUHMWPE samples

material stiffness, it is sufficient to deduce the stroke of the top plate until 10% of the sample height.

5.2   Summary of Chapter Five

This chapter covered three main stages of methodology for this research. The Flow Chart of Methodology provided a schematic overall view of this project with the aim to improve the performance of THR in the aspect of acetabular components. It is important that the methodology conducted achieves the objectives intended for this research.

Reference

1. A.E. Bowden, E. Oneida, J. Bergström, Computer modeling and simulation of UHMWPE. in *UHMWPE Biomaterials Handbook* (2009), pp. 519–532

Chapter 6
Analytical Results of Safe Zone Orientation

6.1 Introduction

This chapter will discuss the results obtained from the three main stage procedures explained in previous chapters. The first part will cover the new safe zone orientation of acetabular cup. The second part exhibits data results from the simulation using FEA. The third part will show the properties of new composite materials and use the data as new input material properties into FEA.

6.2 Analytical Results of Safe Zone Orientation

The results obtained from the workspace tab in MATLAB were copied and transferred into Excel Workbook. There are five types of acetabular components used for the analytical process. The head-neck ratios of 1.83, 2.33, 2.67, 3.0 and 3.33 are used to define the safe zone area for acetabular components. The black area hatched represents the predicted areas for the safe zone orientation in every graph shown in this chapter. The only exception is the case of 1.83 head-neck ratio which has no hatch area highlighted. The FL110 represents the Flexion (FL) of 110°, EXT30 represents the Extension (EXT) of 30°, ER40 represents the External Rotation (ER) of 40°, IR120 represents the Internal Rotation (IR) of 120°, ABD30 represents the Abduction (ABD) of 30° and ADD40 represents the Adduction (ADD) of 40°. All the formulas that have been used for this calculation are intended to find the angles of the six basic motions of the human body which are defined in terms of inclination angle and anteversion angle. The graph of every motion is drawn with their own respective symbols as per agreed from the parameters that have been set before in Sect. 3.2.

Figures 6.1, 6.2, 6.3, 6.4 and 6.5 show the graph results of the five types of acetabular components that had been calculated. The rectangular shape range between 30°

© The Author(s), under exclusive license to Springer Nature Singapore Pte Ltd. 2024   59
M. F. b. A. Manap et al., *Total Hip Replacement (THR)*,
SpringerBriefs in Applied Sciences and Technology,
https://doi.org/10.1007/978-981-96-0975-8_6

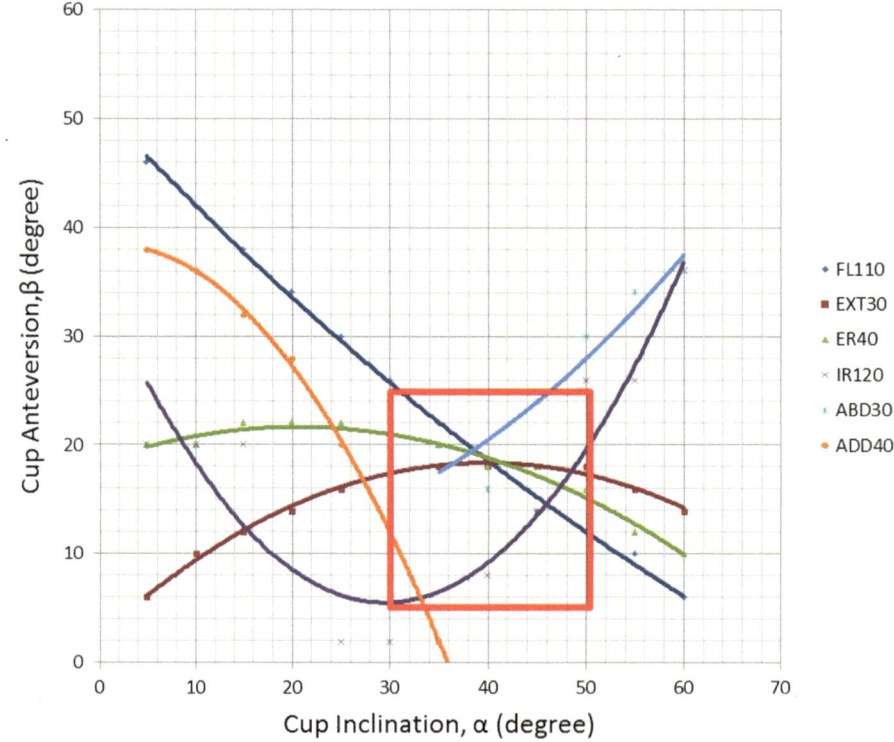

Fig. 6.1 1.83 head-neck ratios which used 22 mm femoral head diameter

and 50° inclination angle combined with 5°–25° anteversion angle is considered as the safe zone orientation defined by Lewinnek et al. in their work. The safe zone was used as the benchmarking for the results in MATLAB as their work was well-known as the pioneer on the introduction of safe zone orientation at the acetabular cup.

From the graphs in Figs. 6.1, 6.2, 6.3, 6.4 and 6.5, a general assumption could be made that using a bigger femoral head size will exhibit a wider safe zone area orientation. The 32, 36 and 40 mm femoral head diameters show a wider shaded area compared to 28 mm femoral head diameter. However, the 22 mm femoral head diameter shows that the range of safe zone orientation does not exist as every plotted graph of the basic six motions does not intersect with each other.

The abduction angle of ABD30 graph shows that in every case of different femoral head size, their limit starts from 30° inclination angle. Based on the formula that was run in MATLAB, putting the inclination angle below 30° with any various anteversion angles will exhibit error as shown in the MATLAB command window tab.

Figure 6.1 shows that for all six basic motions that were intended, a safe zone orientation could not be achieved when dealing with 22 mm femoral head size. The ABD30 graph starts from 35° inclination angle, and any angle below 35° resulted in error value output at the command window during the calculation in MATLAB.

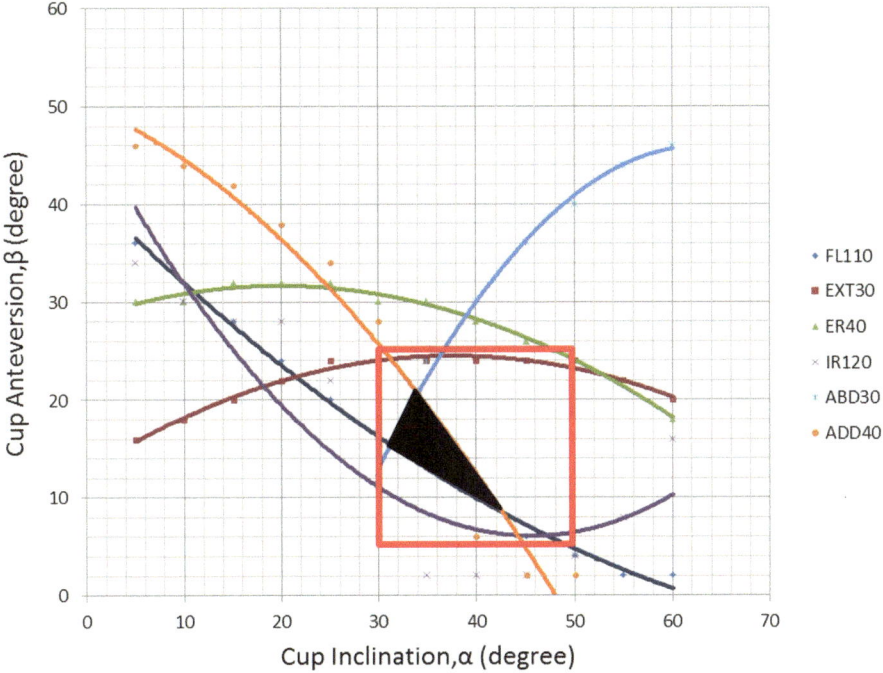

Fig. 6.2 2.33 head-neck ratios which used 28 mm femoral head diameter

Besides, the plot of ADD40 graph is too far away from other parameters; thus a combination area of all the six basic motions could not be achieved. As no hatched area could be found in these graphs, identification of safe zone orientation that is applicable for 22 mm femoral head could not be achieved.

Next, another femoral head size with the same parameters of six basic motions intended will be discussed. Figure 6.2 shows the case of using 28 mm femoral head size which makes 2.33 head-neck ratios. The compliant cup positioning is the area hatched with the maximum value of inclination angle which was within the safe zone of 42° and the minimum inclination angle of 32°. Meanwhile, the maximum anteversion angle is at 20°, and the minimum anteversion angle is at 10°. However, the range of allowable inclination angle is not compulsory to match any range of anteversion angle. It is necessary to refer to the hatched area that has been drawn upon the calculation of the safe zone orientation.

Figure 6.3 shows the case of using 32 mm femoral head diameter with the ratio head-neck approximately at 2.67. The graph exhibits a greater hatched area compared to 28 mm femoral head size diameter. The maximum allowable inclination angle is at 48°, and the minimum allowable inclination angle is at 30°. For the anteversion angle, the maximum value is at 28° and the minimum is at 6°. The data also shows similarities with 28 mm femoral head in which the range of the safe zone must be considered based upon the hatched area.

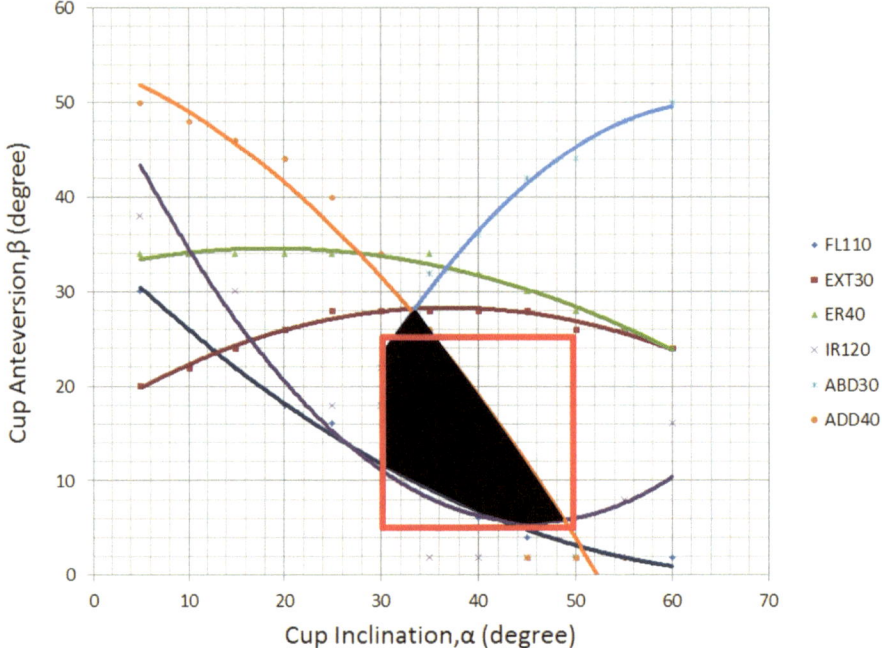

Fig. 6.3 2.67 head-neck ratios which used 32 mm femoral head diameter

Meanwhile, Fig. 6.4 shows the case of using 36 mm femoral head diameter with a 3.0 head-neck ratio. The graph shows almost the same size of hatched area compared to 32 mm femoral head but different from 28 mm femoral head diameter. The maximum inclination angle is at 48°, and the minimum is at 30°. Compared to 28 mm femoral head, the anteversion angle maximum value is greater in which the maximum is at 30° and the minimum is at 4°. However, there was a slight reduction of 2° at the maximum allowable inclination angle when compared to 32 mm femoral head diameter.

The last part of these analytical results is the graph using 40 mm femoral head diameter which makes up the head-neck ratio of approximately 3.33. The hatched area based on Fig. 6.5 shows that the area was almost similar to the case of 32 and 36 mm femoral head size diameter. There is no significant increase in the area hatched as the angles that make out the safe zone orientation do not exhibit out-of-range angles. The maximum inclination angle is at 50°, and the minimum inclination angle is at 30°. On the other hand, anteversion angle maximum value could be seen at 32° and the minimum is at 2°.

The results from this analytical approach agreed with the safe zone orientation introduced by Lewinnek et al. [1] in which the hatched area does not exceed the 50° inclination angle. However, this study showed a narrower range of safe zone orientation compared to their works. This could be due that this work is mainly pivotal on the six basic range motions for daily living activities (ADL) but their work

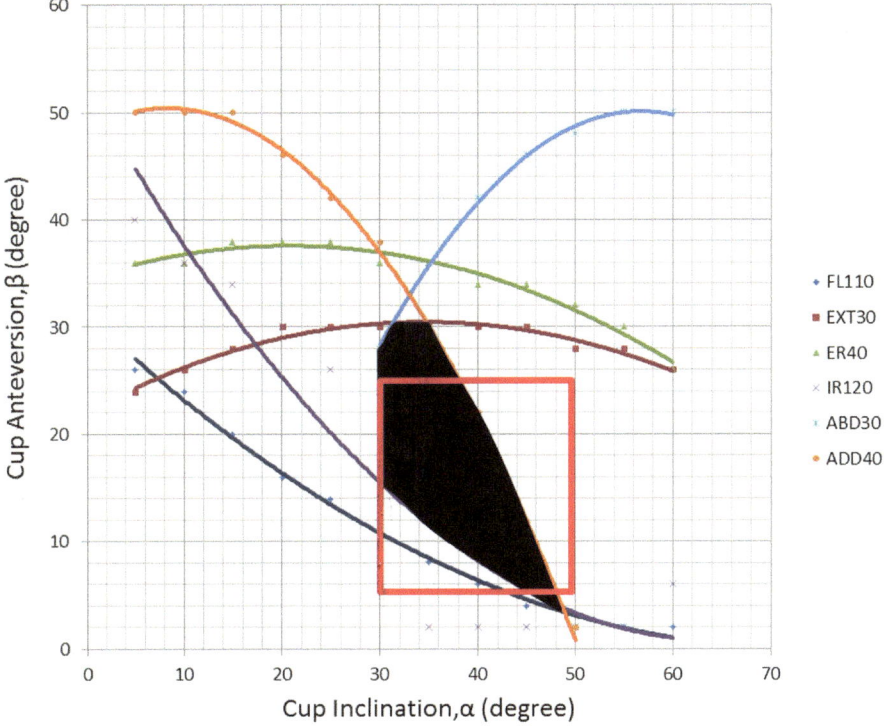

Fig. 6.4 3.0 head-neck ratios which used 36 mm femoral head diameter

used a different approach based on the importance of the acetabular cup position relativity to the body's axis as the reference. In addition, they claimed that the range of safe zone orientation is based on 300 patient cases by using Fisher exact test which is different from this approach.

Based on Chap. 2, the result for the safe zone is contrary to many researchers. Although the research study tried to resolve the inconsistency by providing the average recommendation [2], this study proved that acetabular component sizing played an important criteria in determining the safe zone orientation. A general range of safe zone orientation could not be set as universal as many parameters should be taken into consideration. We tried to minimize other factor consideration such as the neck-stem angle or the design of acetabular components, yet the results indicated that in order to comply with the ROM intended, the safe zone orientation varies. The variations are mainly due to the femoral head diameter in which the results suggested that bigger femoral head diameter will increase the safe zone orientation.

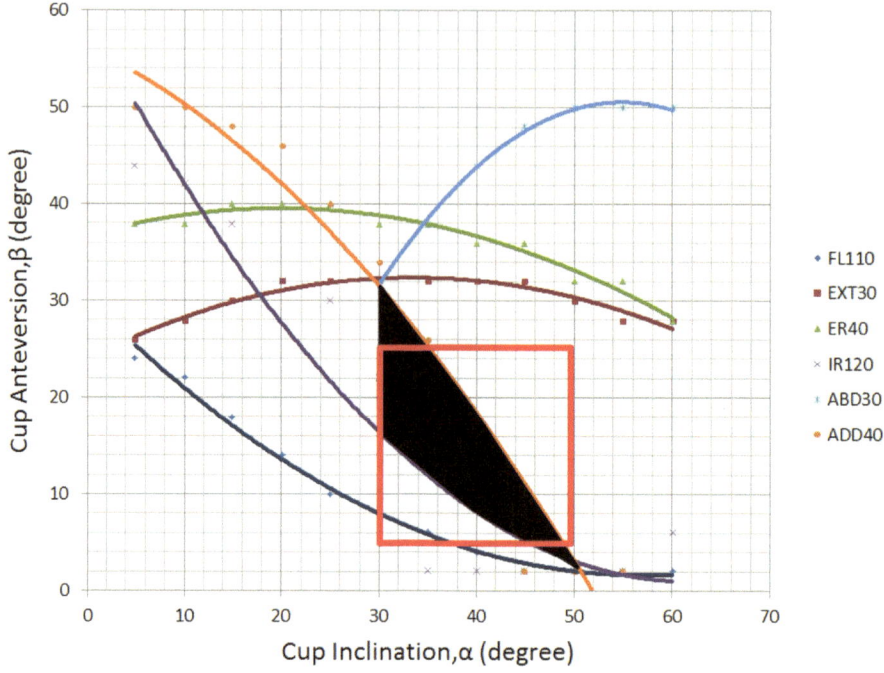

Fig. 6.5 3.33 head-neck ratios which used 40 mm femoral head diameter

6.2.1 The Maximum Safe Inclination Angle

The data from numerical analytical results (Sect. 6.2) shows that by using the 28 mm femoral head diameter, the maximum angle that will allow the safe zone of the acetabular components will make up about 42° inclination angle matched by 10° anteversion angle approximately [3]. Although a greater inclination angle is allowed during the simulation, in the case of 28 mm femoral head, these angles are defined as the maximum that will halt the dislocation and excessive contact pressure on the superior region. However, the angles beyond the analytical results are still important as the focus was on the effect of the edge-loading and contact pressure inside the acetabular region.

For the case of 32 mm femoral head diameter conducted in MATLAB, the analytical data result shows that the maximum safe zone range orientation is at 48° inclination angle combined with 6° anteversion angle. The maximum orientation is particularly for the usage of 32 mm femoral head only. Meanwhile, the 36 mm femoral head diameter will give out the maximum safe zone orientation angle of 48° inclination angle with 4° anteversion angle. Table 6.1 shows the definition of Maximum Safe Inclination Angle acquired from numerical approach using MATLAB.

Table 6.1 The definition of the maximum safe inclination angle with corresponding anteversion angle from analytical approach conducted in the MATLAB

Maximum safe angle	Inclination angle(°)	Anteversion angle(°)
28 mm femoral head	42	10
32 mm femoral head	48	6
36 mm femoral head	48	4

References

1. G. E. Lewinnek, J. L. Lewis, R. Tarr, C. C.L, J. Zimmerman, Dislocations after total hip replacement arthroplasties. J. Bone Joint Surg. Am. **60**, 217–220 (1978)
2. Y.-S. Yoon, A. J. Hodgson, J. Tonetti, B. a Masri, C. P. Duncan, Resolving inconsistencies in defining the target orientation for the acetabular cup angles in total hip arthroplasty. Clin. Biomech. (Bristol, Avon) **23**(3), 253–259 (2008)
3. M Manap S Shuib A Romli A Shokri 2015 The influence of femoral ball size on the range of motion that fulfills the criteria of safe zone orientation acetabular cup Jurnal Teknologi 76 7 31 35

Chapter 7
Finite Element Analysis (FEA) Static Structural Analysis

7.1 FEA Static Structural Analysis Results

Simulation works were conducted by using ANSYS WORKBENCH based on the results collected from the analytical model analysis. There is an intention to consider various femoral head diameter sizes to analyze three critical mechanical aspect assessments from the simulation which are the contact pressure, Von-Mises stress, and total deformation. However, there are two main issues that arise from the simulation which halted us from using too many various acetabular component sizes. Firstly, the parameter consideration was on the Metal-on-Polymer (MoP) THR only. If the femoral head is too big, it will require a greater size of acetabular cup liner and metal backing. Thus, considering the typical hip pelvis of any normal human body, any size of femoral head diameter exceeding 36 mm is not suitable to be used at the THR for the MoP case study.

For the second one, after doing the analytical analysis on defining the safe zone orientation, it was figured out that some sizes are not suitable for the simulation as there is no clear safe zone range for that particular type combination of acetabular components. For instance, based on our analysis in MATLAB from Fumihiro equations [1], it is clearly illustrated that the head-neck ratio of 1.83 (22 mm femoral head) does not show any hatched area that was considered as the safe zone orientation for the acetabular components combination. The graphs developed for the six basic motions should at least intersect with each other; thus a proper safe zone could be defined.

The simulations were done with three main sizes of femoral head diameter which are 28 mm, 32 mm and 36 mm, respectively. These three types of femoral head diameter are chosen from the graph results in the analytical work models. A significance increment of hatched area could be seen during the transition from 28 mm femoral head to 32 mm femoral head diameter. On the other hand, the transition from 32 mm femoral head diameter to 36 mm femoral head diameter does not exhibit a wider hatched area of safe zone orientation. In the current state trend, the majority of

© The Author(s), under exclusive license to Springer Nature Singapore Pte Ltd. 2024
M. F. b. A. Manap et al., *Total Hip Replacement (THR)*,
SpringerBriefs in Applied Sciences and Technology,
https://doi.org/10.1007/978-981-96-0975-8_7

normal procedures involving the THR is considering the femoral head diameter of 28 mm for the replacement at MoP, but our assumption shows that any size greater than 32 mm femoral head diameter should be the priority based on the analytical data observation results that have been made. A note should be taken into account that these assumptions vastly emphasized on the safe zone orientation acetabular components that avoid the dislocation and recurrent impingement at the hip pelvis.

Referring to the second objective, there are three analysis criteria that were planned to achieve by simulation approach; which are the contact pressure, Von-Mises stress and total deformation. The Maximum Safe Inclination Angle is defined as the maximum angle allowable of every type of acetabular components. In detail, the 28 mm femoral head maximum angle is at 42° inclination angle combined with 10° anteversion angle and 32 mm femoral head maximum angle is at 48° inclination angle combined with 6° anteversion angle. On the other hand, the maximum size which is 36 mm femoral head maximum angle is at 48° inclination angle combined with 4° anteversion angle.

7.1.1   Validation FEA by Benchmarking and Hertzian Contact Theory

To validate that our work was done correctly based on enormous research work that had been done in the simulation aspect, a conclusion was deduced by trying to benchmark the work done by Korduba et al. [2] and comparing the total deformation results with Hertzian Contact Theory. This research aims to replicate the work by using the 28 mm femoral head diameter corresponding to the various inclination angle of 0°, 40°, 50° and 70°, respectively. The results agreed with Korduba et al. works as the value was less than 12% errors compared to their results in terms of contact pressure distribution inside the acetabular cup. Our results show that all the contact pressure data recorded exhibited lesser value at 0°, 50° and 70° as compared to their work. However, the data from our imitation at 40° shows a greater value of contact pressure of approximately 0.025 MPa as compared to their work (Fig. 7.1).

On the other hand, calculation based on Hertzian Contact Theory equation on finding the displacement, δ and comparing it with FEA results was also conducted in this research. The data shows the error produced was about 11.3% compared to the 36 mm femoral head at inclination of 0°. Regardless, it is assumed that our simulation methodology is put into a proper way; thus the addition of parameters in this study that was run in ANSYS WORKBENCH V15 is considered valid.

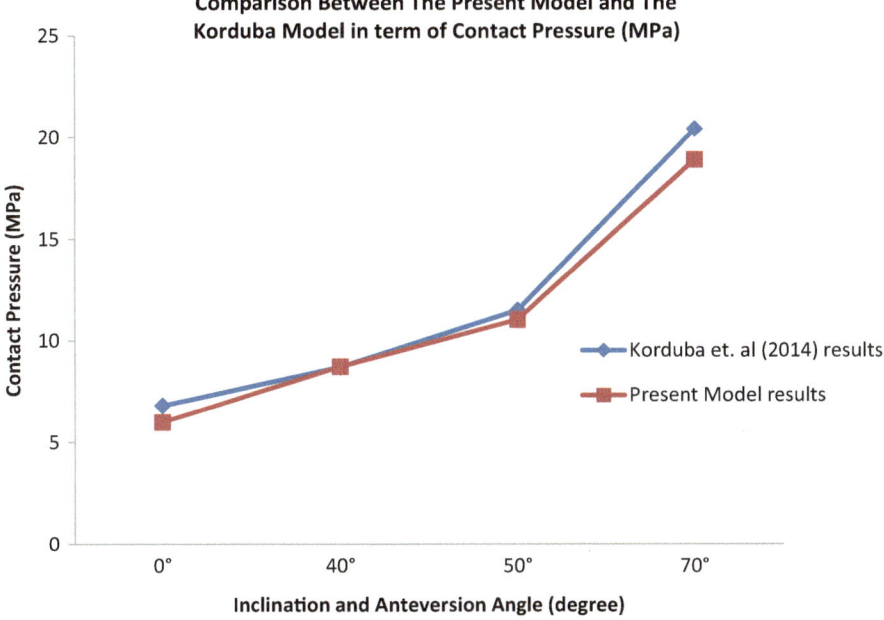

Fig. 7.1 Comparison between the present model and the Korduba model in predicting the contact pressure of acetabular cup for validation purpose

7.1.2 The Mechanical Analysis Results from FEA

Figure 7.2 shows a trend at all orientations intended, and the value of contact pressure decreases when corresponding to bigger femoral head diameter. The orientation of acetabular components is important since most of the previous studies considered that the maximum allowable inclination angle is at 50° [3]. Thus, based on the contact pressure graph, the inclination of 50° shows a significant reduction of contact pressure of about 38.7% which could be seen during the interchange from 28 to 32 mm femoral head diameter. Meanwhile, replacing 32–36 mm femoral head diameter only exhibits about 21.4% reduction of contact pressure inside the acetabular cup at the inclination angle of 50°. However, in the case of considering transition directly from 28 to 36 mm femoral head diameter, the reduction percentage is more than half which is approximately at 51.8%.

At the Maximum Safe Inclination Angle, the contact pressure was reduced 11.2% when changing it from 28 to 32 mm femoral head diameter. Meanwhile, changing 32–36 mm femoral head will reduce contact pressure at 21.9%. Direct transition from 28 to 36 mm femoral head will reduce 30.7% of contact pressure.

Figure 7.3 shows the graph of Von-Mises stress distribution at the acetabular cup. The trend also looks similar with the contact pressure from the angle between 0° and 70°, but there is a slightly different trend if the anteversion angle is included at the

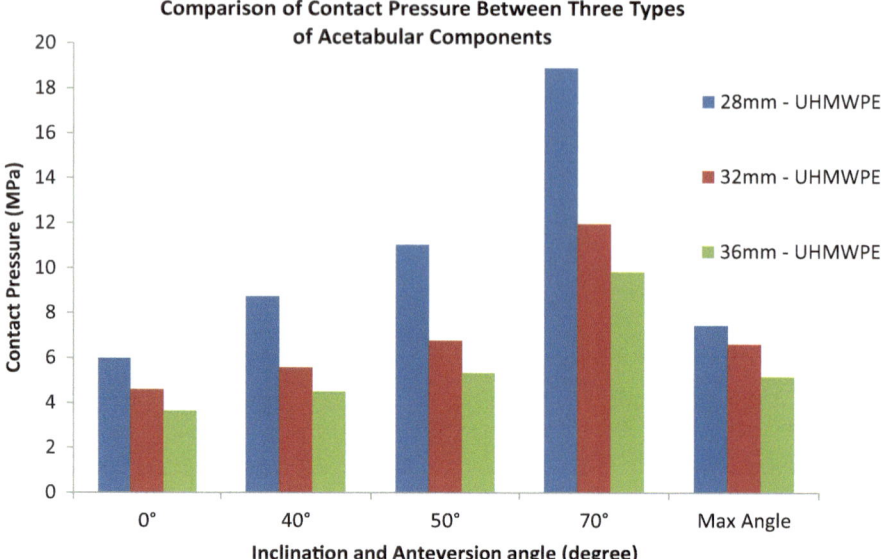

Fig. 7.2 Comparison contact pressure distribution between three types of acetabular components based on various orientations. Note: Maximum Safe Inclination Angle is referred to the maximum angle allowable of every type of acetabular components

Maximum Safe Inclination Angle. A lower Von-Mises stress is better for any design in the engineering system which should not exceed the material Yield Strength. At 50° inclination angle, replacing from 28 to 32 mm femoral head diameter, the Von-Mises stress decreases about 31% and on the other hand, replacing 32–36 mm femoral head diameter will reduce the Von-Mises stress at about 14.6% only. Direct transition from 28 to 36 mm will produce 41.1% reduction of Von-Mises stress.

However, at the Maximum Safe Inclination Angle, the Von-Mises stress shows a slight increment when it transits from 28 to 32 mm femoral head diameter with 5.8%. Yet, the transition from 32 to 36 mm femoral head diameter will reduce the Von-Mises stress at 13.9% and direct transition from 28 to 36 mm femoral head diameter will reduce Von-Mises stress at 8.9%.

The Maximum Safe Inclination Angle shows that using 32 mm femoral head diameter at 48° inclination angle combined with 6° anteversion angle shows the greatest Von-Mises stress of about 7.46 MPa. Nonetheless, 28 mm femoral head diameter at its Maximum Safe Inclination Angle of 42° inclination angle combined with 10° anteversion angle shows a slight lower Von-Mises stress of 7.05 MPa. Overall, the bar graph comparison of Maximum Safe Inclination Angle bar chart confirmed that 36 mm femoral head diameter displayed the lowest Von-Mises stress with a reduction of about 13.9% when compared to the highest Von-Mises stress which is at 32 mm femoral head diameter. It is believed that the effect of anteversion angle resulted in higher Von-Mises stress of 32 mm femoral head diameter compared to 28 mm femoral head diameter. The Yield Strength of typical UHMWPE is about

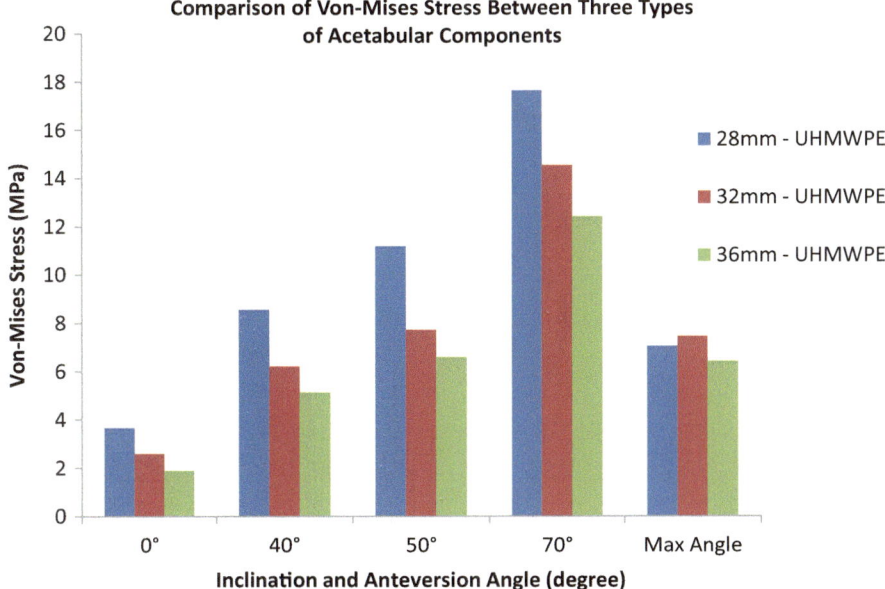

Fig. 7.3 Comparison of Von-Mises stress between three types of acetabular components

20 MPa or more [4, 5] and the results of all orientation angles do not exceed 20 MPa at the 2450 N loading condition. A deduction could be made that the design in all types of acetabular components is safe upon all orientations intended based on FEA results. However, consideration of any higher loading condition such as climbing stairs activity or stumbling condition could induce implant failure, especially using 28 mm femoral head diameter at 70° inclination angle. This is due to that stumbling condition that could reach up to eight times of body weight which may consequence with higher Von-Mises stress that exceeds the UHMWPE limit.

Figure 7.4 shows the scattered bar of total deformation of every type of acetabular components with the same conditions applied. In this study, the data exhibits that using 28 mm femoral head diameter at 70° inclination angle will expose the greatest deformation of 0.1377 mm at the superior region of the cup compared to other results. Meanwhile at 50° inclination angle, the total deformation reduction from 28 to 32 mm femoral head shows about 44.1%, and it is regarded as the highest deformation downsizing upon the transition from 28 to 32 mm femoral head diameter. Transition from 32 to 36 mm will produce reduction of total deformation at 17.4% and direct transition from 28 to 36 mm will produce reduction of total deformation at 53.9%.

The Maximum Safe Inclination Angle data shows that transition from 28 to 32 mm and 32–36 mm femoral head diameter will reduce the total deformation at 4.1% and 17.9%, respectively. On the other hand, transition from 28 to 36 mm femoral head will reduce the total deformation at 21.3%.

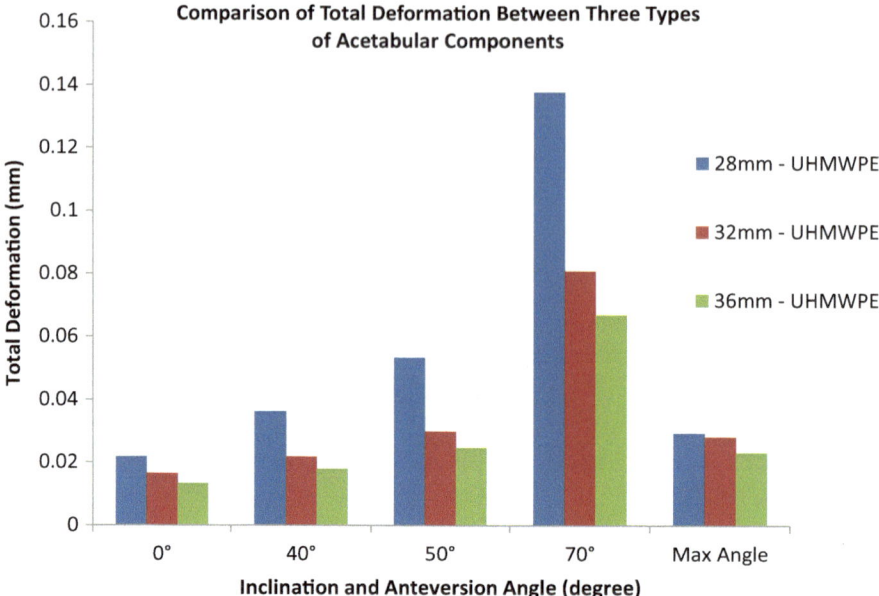

Fig. 7.4 Comparison of total deformation between three types of acetabular component

The data result that compares the 50° inclination angle (which is considered as the maximum inclination angle from many researchers) with our Maximum Safe Inclination Angle for the 28, 32 and 36 mm indicates a slight reduction. The results show that Maximum Safe Inclination Angle of 28 mm, 32 mm and 36 mm femoral head diameter exhibits lower total deformation with a percentage of 45.1%, 5.6% and 6.2%, respectively.

The results in this section show that the Maximum Safe Inclination Angle exhibits significant reduction as compared to 50° inclination angle at 28 mm femoral head diameter. The summarization of the reduction in all three types of acetabular components is shown in Table 7.1. The reduction values of 28 mm femoral head diameter at Maximum Safe Inclination Angle instead of 50° inclination angle for contact pressure, Von-Mises stress and total deformation were 32.69%, 37.05% and 45.05%, respectively. The 32–36 mm femoral head diameter cases show reduction range of less than 10% in all three analyses conducted in FEA. General observation could deduce that the Maximum Safe Inclination Angle orientation performs better compared to 50° inclination angle. It is believed that the corresponding anteversion angle of Maximum Safe Inclination Angle resulted in lowered mechanical failure analysis data. The Maximum Safe Inclination Angle of 28 mm femoral head diameter only exhibits the inclination of 42°, but the corresponding anteversion angle of 10° proved an important criteria based on the FEA results. The higher anteversion angle resulted in lower contact pressure, Von-Mises stress and total deformation.

Table 7.1 The reduction percentage of mechanical analysis in FEA at Maximum Safe Inclination Angle

Parameters	Analysis	Reduction percentages (%)		
		28 mm	32 mm	36 mm
50° inclination change to maximum safe inclination angle	Contact pressure	32.69	2.53	3.19
	Von-Mises stress	37.05	3.47	2.61
	Total deformation	45.05	5.62	6.21

However, consideration of safe zone orientation should be taken into account in which the anteversion angle must be in accordance with the limit of the safe zone.

7.1.3 Contour Analysis for Edge-Loading Effect

This subsection focuses on the contour analysis of the three types of acetabular components that had been simulated in the ANSYS WORKBENCH. Figure 7.5 represents the contour comparison in terms of contact pressure between the three types of acetabular components based on their dimension. The contact areas are inside the acetabular cup as the force is being exerted from the femoral head size. The contact is different upon different orientation angles and different femoral head size diameters. Two different orientations were tabulated in this study which are 50° inclination angle and the maximum angle. The highest contour patch (in red color) clearly extends to the rim of the acetabular cup that represents the superior region. Meanwhile, the Maximum Safe Inclination Angle does not show the contour patches exceed the rim, particularly from 32 and 36 mm femoral head diameter. The definition of edge-loading in MoP THR condition is defined as the highest contact patch extended to the rim of the acetabular cup [6].

These contours displayed the trend that higher inclination angle will cause the contact pressure to move toward the superior region of the acetabular cup. Even though different types of acetabular components were simulated in ANSYS, but in general, the trends look similar; thus it agrees that higher inclination angle will induce higher contact pressure inside the acetabular cup. However, the analyzed data is being represented in Fig. 7.6 at inclination angle of 40°, and the marking contour patch is clearly different with respect to femoral head size diameter. The contour patch areas of 28 mm femoral head diameter is at the highest on the upper part of the superior cup region (as marked in red color); conversely, 32 mm and 36 mm femoral head diameter shows the highest contour patch on a slightly lower part of the superior region acetabular cup.

Meanwhile, observation shows that the contour of contact pressure is much wider on 32 and 36 mm femoral head compared to 28 mm femoral head diameter (as marked in black rectangular shape). From the observation, assumption could be made that a bigger femoral head diameter of more than 28 mm will increase the articulate

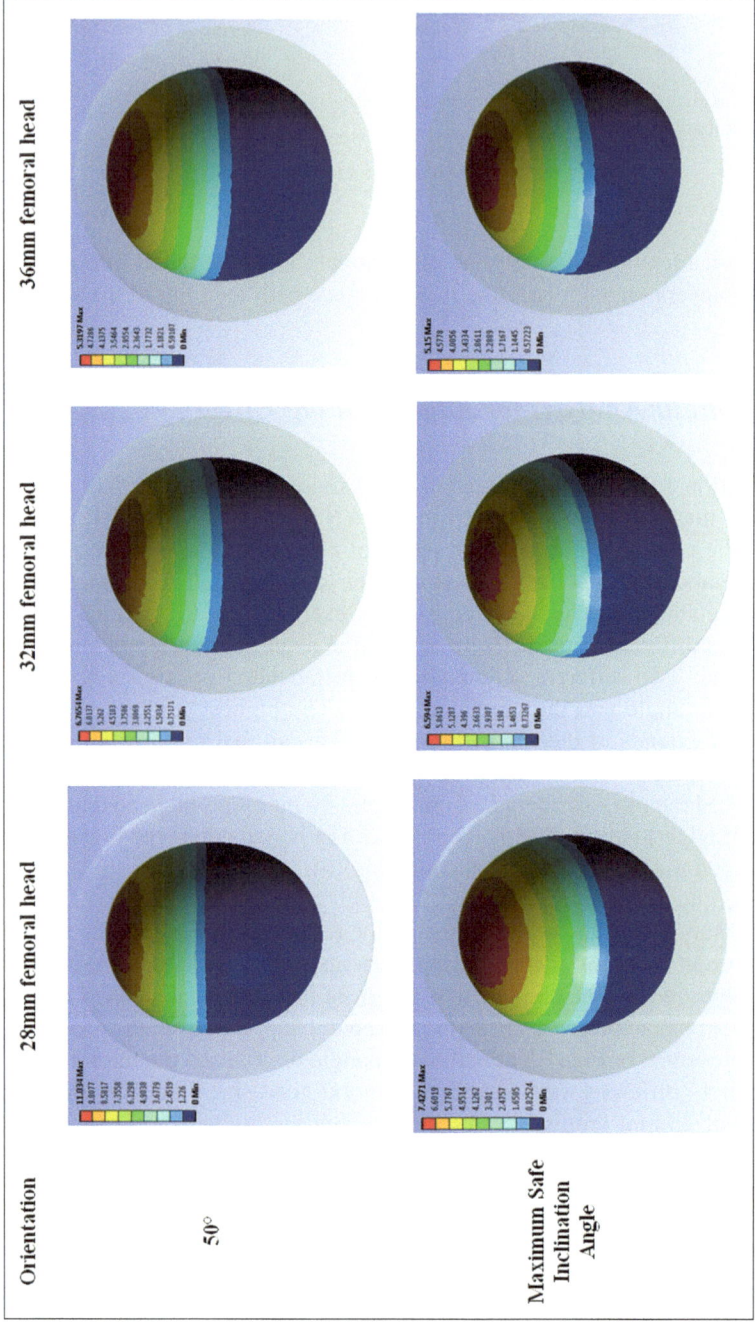

Fig. 7.5 The contour comparison of all types of acetabular components at 50° inclination angle versus Maximum Safe Inclination Angle

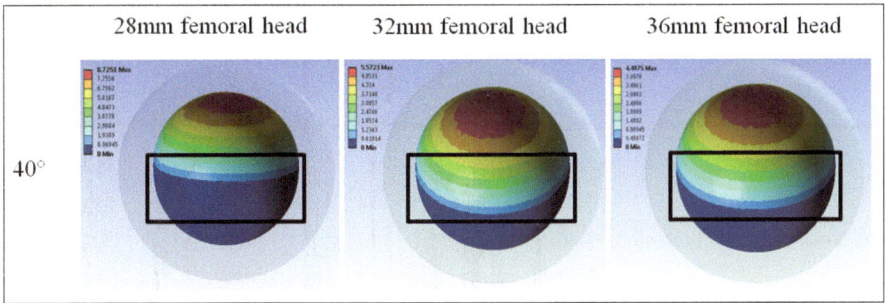

Fig. 7.6 Comparison contour analysis of three types of acetabular component at 40° inclination angle

contact area between the femoral head and acetabular cup. Consequently, it will reduce the maximum contact pressure inside the acetabular cup particularly at the superior region.

This study used only average gait cycle of normal walking neglecting other gait cycle such as stair climbing, standing up and knee bending. The author knows that ADL may affect the edge-loading effect as retrieval CoC THR shows that climbing stairs or rising from stairs may cause greater edge-loading compared to other motion [7]. However, this study focuses on the orientation and types of the acetabular cup; thus a constant loading force is needed for comparison purpose.

Previous studies have shown that edge-loading effect at any angle beyond 55° cup inclination was predicted during normal walking, ascending and descending stairs activities [6]. The present study which is on normal walking mode supported this conclusion based on the contour patch extension on the rim observation that has been shown. Despite that, the edge-loading effect could be seen starting from 50° inclination angle in this project for the three types of acetabular components. The inclination angle 70° exhibits total edge-loading at the superior region of the cup where the contact patch extends over the rim of the acetabular cup. On the other hand, the Maximum Safe Inclination Angle of the three types of acetabular components does not show any sign of edge-loading at the superior region. Thus, an agreement could be deduced that the safe zone predicted for the respective types is suitable and considered as the maximum allowable orientation for them.

Orientation played an important role as inclination angle at 70° should not be used as a consideration of any type of acetabular component sizes. Meanwhile, the author does believe that inclination angle of 50° is not suitable on certain femoral head size, especially on 28 mm femoral head diameter. Previous studies exhibit that edge-loading can be avoided at inclination angle less than 50° radiographically [8, 9]. Meanwhile, other studies have shown that inclination angle no larger than 45° is best for achieving stability, avoiding dislocation and preventing wear [10, 11]. We could confirm that inclination of less than 50° could avoid edge-loading based on our numerical approach (safe zone hatched) that does not even reach 50° inclination angle for 28, 32 and 36 mm femoral head diameter. Contrary to studies by [10, 11],

Table 7.2 Summary mechanical engineering analysis based on FEA results

Orientation (°)	Analysis	Reduction percentages (%)		
		28–32 mm	32–36 mm	28–36 mm
50	Contact pressure	38.7	21.4	51.8
	Von-Mises stress	31.0	14.6	41.1
	Total deformation	44.1	17.4	53.9
Maximum safe inclination angle	Contact pressure	11.2	21.9	30.7
	Von-Mises stress	− 5.8	13.9	8.9
	Total deformation	4.1	17.9	21.3

our study shows that angle of more than 45° could still be used for 32 and 36 mm femoral head size with matched specific anteversion angle included.

Table. 7.2 shows the key finding that had been summarized based on the FEA analysis in Chapter Four. The overall data exhibits that reduction of contact pressure, Von-Mises stress and total deformation was recorded when using bigger femoral head diameter. It is believed that Maximum Safe Inclination Angle orientation influences the percentage reduction of the analysis results. This finding was based on the Von-Mises stress analysis recorded during transition from 28 to 32 mm femoral head diameter which shows a slight increment of 5.8%.

References

1. F Yoshimine K Ginbayashi 2002 A mathematical formula to calculate the theoretical range of motion for total hip replacement J. Biomech. 35 7 989 993
2. LA Korduba A Essner R Pivec P Lancin MA Mont A Wang RE Delanois 2014 Effect of acetabular cup abduction angle on wear of ultrahigh-molecular-weight polyethylene in hip simulator testing Am. J. Orthop. 43 10 466 471
3. H Miki T Kyo Y Kuroda I Nakahara N Sugano 2014 Risk of edge-loading and prosthesis impingement due to posterior pelvic tilting after total hip arthroplasty Clin. Biomech. (Bristol, Avon) 29 6 607 613
4. D Kluess H Martin W Mittelmeier K-P Schmitz R Bader 2007 Influence of femoral head size on impingement, dislocation and stress distribution in total hip replacement Med. Eng. Phys. 29 4 465 471
5. G Saxler A Marx D Vandevelde U Langlotz M Tannast M Wiese U Michaelis G Kemper PA Grützner R Steffen M Knoch von T Holland-Letz K Bernsmann 2004 The accuracy of free-hand cup positioning—a CT based measurement of cup placement in 105 total hip arthroplasties Int. Orthop. 28 4 198 201
6. X Hua J Li Z Jin J Fisher 2016 The contact mechanics and occurrence of edge loading in modular metal-on-polyethylene total hip replacement during daily activities Med. Eng. Phys. 38 6 518 525
7. C. I. Esposito, W. L. Walter, A. Roques, M. A. Tuke, B. A. Zicat, W. R. Walsh, W. K. Walter, Wear in alumina-on-alumina ceramic total hip replacements: a retrieval analysis of edge loading. J. Bone Jt. Surgery. Br. Vol. **94**(7), 901–907 (2012)

8. M. C. Callanan, B. Jarrett, C. R. Bragdon, D. Zurakowski, H. E. Rubash, A. A Freiberg, H. Malchau, The John Charnley Award: risk factors for cup malpositioning: quality improvement through a joint registry at a tertiary hospital. Clin. Orthop. Relat. Res. **469**(2), 319–329 (2011)
9. N. J. Little, C. A Busch, J. A Gallagher, C. H. Rorabeck, R. B. Bourne, Acetabular polyethylene wear and acetabular inclination and femoral offset. Clin. Orthop. Relat. Res., **467**(11), 2895–900 (2009)
10. S Patil A Bergula PC Chen CW Colwell Jr DD D'Lima 2003 Polyethylene wear and acetabular component orientation J. Bone Joint Surg. Am. 85 4 56 63
11. R. P. Robinson, P. T. Simonian, I. M. Gradisar, R. P. Ching, Joint motion and surface contact area related to component position in total hip arthroplasty. J. Bone Jt. Surgery. Br. Vol. **79**(1), 140–146 (1997)

Chapter 8
Experimental Work of EPUHMWPE Materials

8.1 Experimental Work of EpUHMWPE Materials

In this subchapter, the new composite variants of Young's Modulus and new simulation study based on the new composite data were shown. The first part is the result of the compression test upon finding the Young's Modulus of each variant and the second part is the simulation comparison based on the new input obtained from the experimental setup as explained in Chap. 8.

8.1.1 Compression Test Results for All Variants

The aim of the compression test was to find the Young Modulus of every variant that will be used as input engineering data in the ANSYS software. This test was conducted given the fact that no engineering data was available for these variants that had been proposed as the replacement of UHMWPE. The results from every variant are tabulated in Table 8.1 with every variant showing a different value and it is noticed that the EpUHMWPE5 gives the highest value of Young Modulus compared to other variants. From the data shown, EpUHMWPE5 recorded the highest Young's Modulus value compared to other variants. EpUHMWPE0 was selected as the controlled variable which shows the value of 1099.76 MPa. As the variant percentages increased, the value of Young's Modulus also increased. However, it will increase only until 5% variants as the data showed that increasing the variant of more than 5% will reduce the Young's Modulus value. The assumption could be made that the variant of 5% is the maximum reinforcement material allowable to achieve the maximum Young's Modulus value.

Before the specimens of each variant were compressed into compression machine, visual inspection was observed on every composition variant. Every specimen was stacked in each variant with the sequence from 0 to 10% as

M. F. b. A. Manap et al., *Total Hip Replacement (THR)*,
SpringerBriefs in Applied Sciences and Technology,
https://doi.org/10.1007/978-981-96-0975-8_8

Table 8.1 Young's Modulus and compression strength value of the variant proposed as a replacement of acetabular cup

Sample variant	Young's modulus (MPa)	Compression strength (MPa)
EpUHMWPE10	1031.49	34.8309
EpUHMWPE7	1188.43	40.0394
EpUHMWPE5	1338.29	44.6123
EpUHMWPE3	1155.38	38.5126
EpUHMWPE1	1049.52	34.9862
EpUHMWPE0	1099.76	36.6593

in Fig. 8.1. EpUHMWPE0 represented a total of 100% Epoxy without incorporated with UHMWPE. Meanwhile, the other five specimen symbols were denoted with EpUHMWPE1, EpUHMWPE3, EpUHMWPE5, EpUHMWPE7 and EpUHMWPE10 that represent the percentage of UHMWPE of 1%, 3%, 5%, 7% and 10%, respectively. EpUHMWPE0, EpUHMWPE1 and EpUHMWPE3 do not exhibit significant existence of UHMWPE powder that has been incorporated into Epoxy matrix. However, dispersion of UHMWPE starts to be seen in EpUHMWPE5. In the case of EpUHMWPE7 specimen, it seems that two layers were formed on that specimen; at the top is the UHMWPE powder and at the bottom is the Epoxy. Even though two layers are obviously formed at EpUHMWPE7, this new composite is still intact without any UHMWPE powder debris coming out from the composition.

Our study is focusing on finding Young's Modulus which is the main critical data needed in order to do the simulation. Thus, the observation via naked eye is only to get the main idea of how the percentage of UHMWPE affects the composition of every variant shown. In addition to that, the author could conclude that different densities of UHMWPE and Epoxy resulted in the layered form at every variant.

8.1.2   Simulation Results Based on New Data

In this part, it is intended to propose a new material composition as the replacement of UHMWPE at acetabular cup. With the new data record of EpUHMWPE variants upon the compression test, the data requirements for the simulation were tabulated as shown in Table 8.2.

The data for the entire variant is used as input engineering data in ANSYS WORKBENCH in comparison with the UHMWPE recorded earlier. As the simulation approach was conducted by using three different acetabular components, the new data materials on the entire three components were re-run in that simulation. The focal analyses mainly depend on the orientation angle which are the 50° inclination angle and the Maximum Safe Inclination Angle that were figured out in Sect. 4. 1. The aim is to compare the results for both orientation angle and 50° inclination

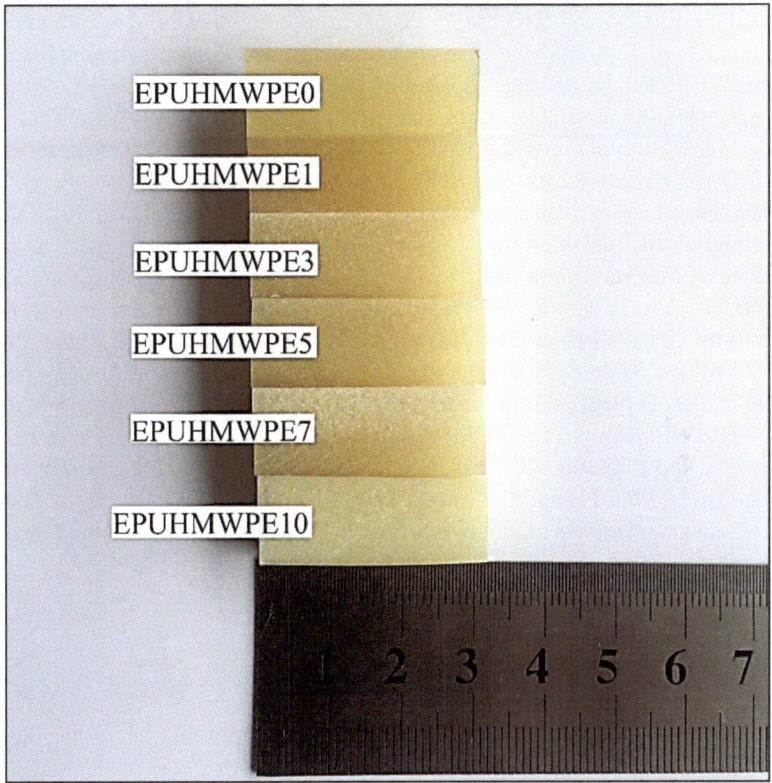

Fig. 8.1 The variant samples observed. Two layers are clearly formed for variant EpUHMWPE5 and EpUHMWPE7

Table 8.2 The input engineering data required upon doing simulation

Sample variant	Young's modulus (MPa)	Poisson's ratio
EpUHMWPE10	1031.49	0.35
EpUHMWPE7	1188.43	0.35
EpUHMWPE5	1338.29	0.35
EpUHMWPE3	1155.38	0.35
EpUHMWPE1	1049.52	0.35
EpUHMWPE0	1099.76	0.35

angle is chosen for the benchmarking purpose as it is the recommendation position of acetabular cup by most researchers [1].

Figure 8.2 shows the histogram comparison of material variations at 50° inclination angle with three types of acetabular components. The 28 mm femoral head cases show that the highest contact pressure recorded by EpUHMWPE5 and the lowest

recorded by EpUHMWPE10 are 11.05 MPa and 10.91 MPa, respectively. Meanwhile, the 32 mm femoral head cases exhibit the highest contact pressure recorded by
EpUHMWPE5 and the lowest recorded by EpUHMWPE10 with values of 6.81 MPa
and 6.67 MPa, respectively. On the other hand, 36 mm femoral head cases show
that the highest contact pressure recorded by UHMWPE and the lowest recorded by
EpUHMWPE10 are the values of 5.32 MPa and 5.22 MPa, respectively.

The other analysis that was intended to highlight is the Maximum Safe Inclination Angle orientation. Figure 8.3 exhibits the histogram of comparison material
variations at the Maximum Safe Inclination Angle with three types of acetabular
components. The 28 mm femoral head case shows that the highest contact pressure
recorded by UHMWPE and the lowest recorded by EpUHMWPE10 are 7.43 MPa
and 7.20 MPa, respectively. At the meantime, the 32 mm femoral head case shows
that the highest contact pressure recorded by EpUHMWPE5 and the lowest recorded
by EpUHMWPE10 are at 6.67 MPa and 6.59 MPa, respectively. Final histogram
comparison shows that the highest contact pressure for 36 mm femoral head case is
recorded by UHMWPE and the lowest is recorded by EpUHMWPE10. The values
for the highest and lowest contact pressure are 5.15 MPa and 4.98 MPa, respectively.

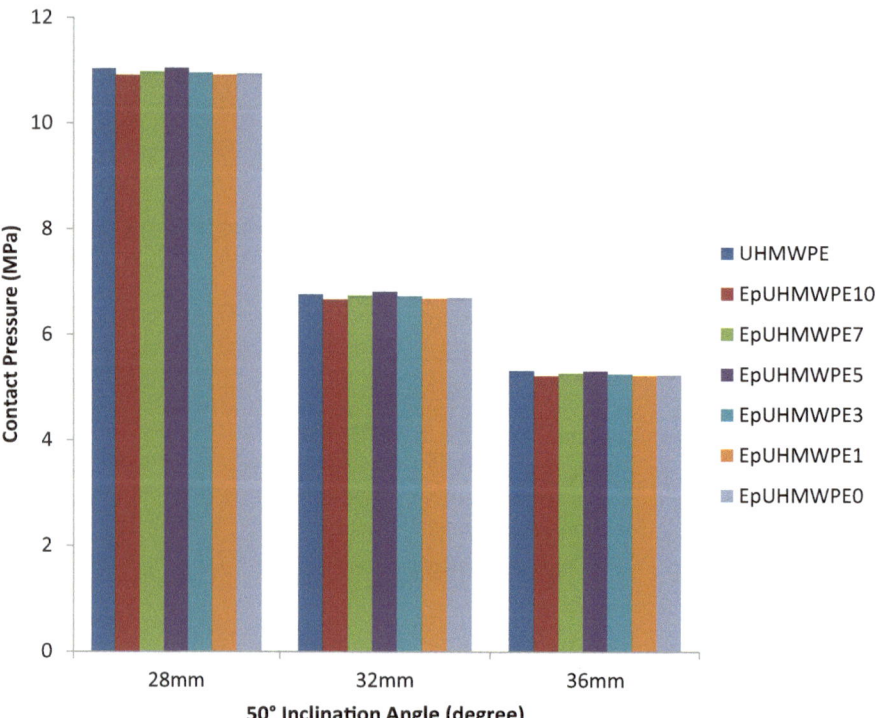

Fig. 8.2 The contact pressure comparison of material variation in three types of acetabular
components at 50° inclination angle

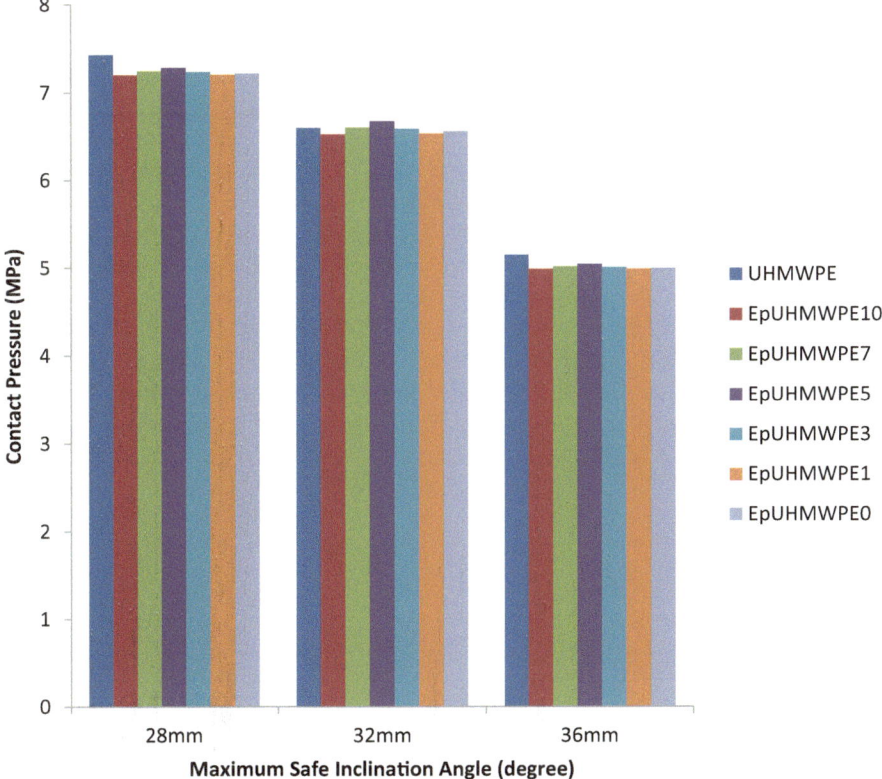

Fig. 8.3 The contact pressure comparison of material variation in three types of acetabular components at the Maximum Safe Inclination Angle

Based on these two orientation angles comparison, a conclusion could be made that EpUHMWPE10 exhibits the lowest contact pressure at any type of acetabular component sizes. Meanwhile, the highest contact pressure varies between UHMWPE and EpUHMWPE5 as among the highest value recorded based upon these two orientation angles. Yet, this analysis showed that bigger femoral head will reduce the contact pressure at the articulation surfaces. A deduction was made that the EpUHMWPE10 recorded the lowest contact pressure due to its lowest Young's Modulus value among other variants.

At 50° inclination angle case study, the EpUHMWPE5 shows a reduction of 0.15% contact pressure compared to existing material of UHMWPE when using 36 mm femoral head diameter. Meanwhile, at the maximum angle case study, the EpUHMWPE5 shows a reduction of 2.16% contact pressure compared to UHMWPE in 36 mm femoral head diameter case. The results show that EpUHMWPE5 reduced the contact pressure inside the acetabular cup as compared to UHMWPE.

The next analysis is in terms of Von-Mises stress distribution. Figure 8.4 shows the histogram comparison of Von-Mises stress with different material variations at

50° inclination angle. The 28 mm femoral head case shows that UHMWPE records the highest Von-Mises stress and the lowest is recorded by EpUHMWPE10 with the values of 11.20 MPa and 10.50 MPa, respectively. Meanwhile, the 32 mm case exhibits that the highest Von-Mises stress is recorded by UHMWPE and the lowest Von-Mises stress is recorded by EpUHMWPE10 with the values of 7.73 MPa and 6.10 MPa, respectively. The final part from this figure shows the 36 mm femoral head case which has the highest and lowest Von-Mises stress recorded by UHMWPE and EpUHMWPE5. Different material variations recorded the lowest Von-Mises stress for 36 mm femoral head case. The values recorded for the highest and lowest Von-Mises stress are 6.59 MPa and 4.68 MPa, respectively.

On the other hand, Fig. 8.5 shows the histogram comparison of Von-Mises stresses with different material variations at Maximum Safe Inclination Angle orientation. The 28 mm femoral head case shows that UHMWPE records the highest Von-Mises stress and the lowest is recorded by EpUHMWPE10 at the values of 7.05 MPa and 6.65 MPa, respectively. Meanwhile, 32 mm femoral head also recorded the same variant of UHMWPE at the highest and EpUHMWPE10 at the lowest with the values

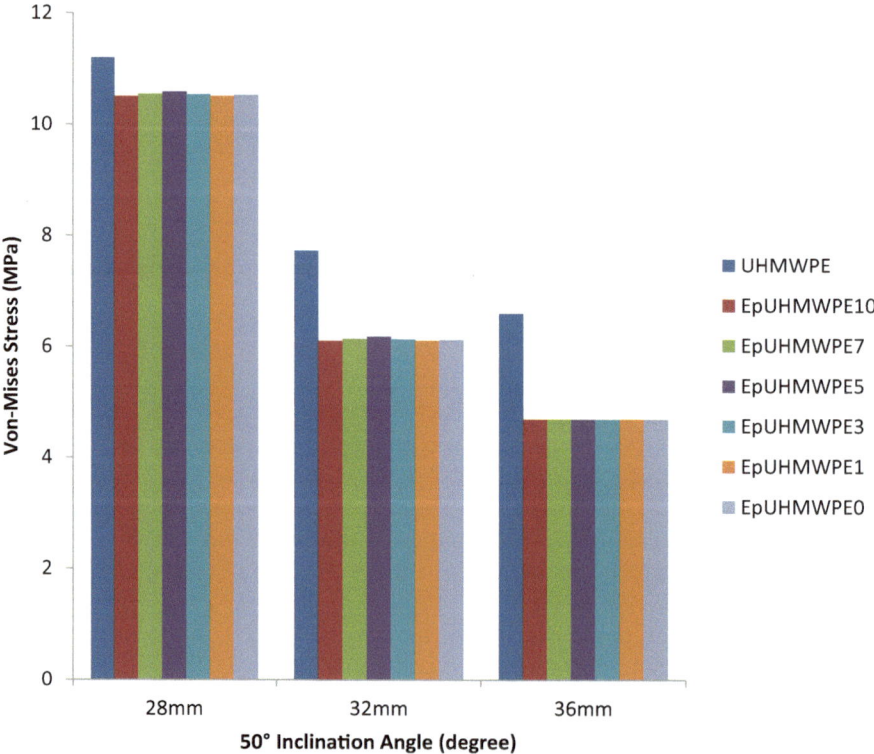

Fig. 8.4 The Von-Mises stress comparison of material variation in three types of acetabular components at 50° inclination angle

of 7.46 MPa and 5.85 MPa, respectively. The final part which is 36 mm femoral head has the highest Von-Mises stress recorded by UHMWPE and the lowest is recorded by EpUHMWPE5 with the values of 6.42 MPa and 4.50 MPa, respectively. On the other hand, the data recorded that transition from 28 to 32 mm femoral head at the Maximum Safe Inclination Angle using UHMWPE exhibits different pattern from our prediction. Increasing the femoral head from 28 to 32 mm will increase the Von-Mises stress at approximately 5.49%. Even though the author could not exactly explain the reason behind this fact, it is believed that orientation (Maximum Safe Inclination Angle) and UHMWPE properties played a vital role to this effect. The Poisson's ratio of UHMWPE close to 0.5 shows that throughout the compression of UHMWPE, it will easily undergo expansion; thus it might affect the Von-Mises stress value.

Meanwhile, other variants analyses show a reduction of Von-Mises stress with higher femoral head diameter size. Although the 32 mm femoral head diameter case shows a higher value compared to 28 mm femoral head diameter in the aspect from Maximum Safe Inclination Angle value, the value is still lower than UHMWPE Yield Strength and 50° inclination angle case. The design is also safe when the analysis

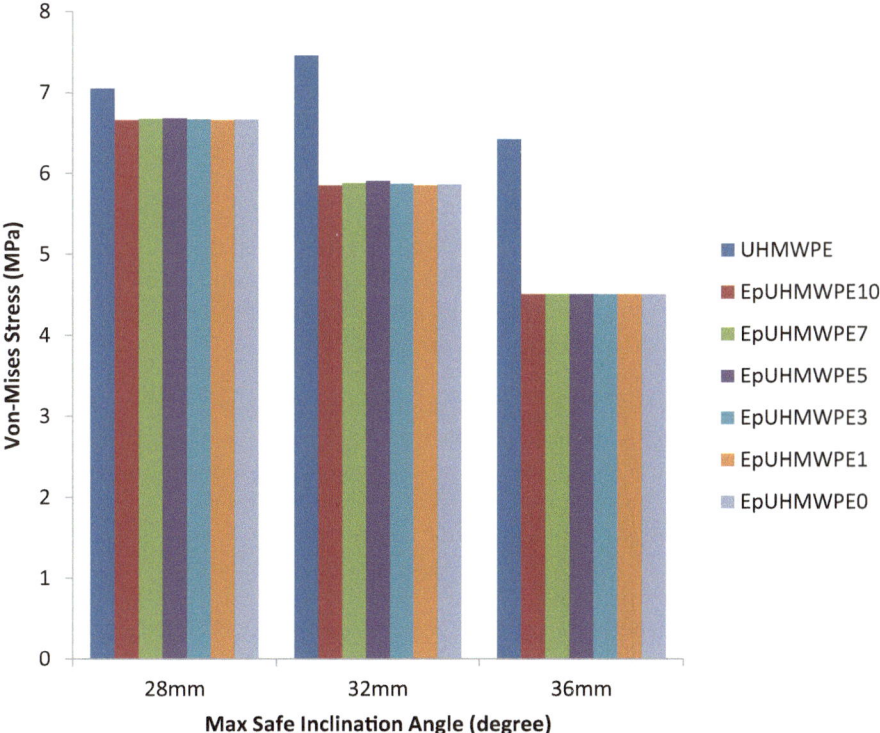

Fig. 8.5 The Von-Mises stress comparison of material variation in three types of acetabular components at the maximum safe inclination angle

was done at Maximum Safe Inclination Angle orientation angle. In addition, the three types of acetabular components analyzed in Maximum Safe Inclination Angle show better results compared to 50° inclination angle in terms of Von-Mises stress Distribution.

General assumption could be made that UHWMPE recorded the highest Von-Mises stress in three different types of acetabular components at these two orientation angles. The highest Von-Mises stress distribution is exhibited by EpUHMWPE10 in both cases of 28 mm and 32 mm femoral head diameter. On the other hand, 36 mm case shows that EpUHMWPE5 recorded the lowest Von-Mises stress.

The total deformation analyses for these two comparisons are shown in Figs. 8.6 and 8.7 which exhibit greater significant difference in between variants compared to contact pressure and Von-Mises stress distribution results. EpUHMWPE5 exhibits the lowest deformation at the superior region of the acetabular cup compared to other variants in all types of acetabular component sizes. At 50° inclination angle, EpUHMWPE5 recorded 0.04181 mm, 0.02327 mm and 0.01915 in 28 mm, 32 mm and 36 mm femoral head size diameter, respectively. Meanwhile, the existing material of UHMWPE shows the second highest total deformation with a value of 0.05329 mm, 0.02976 mm and 0.02456 mm in 28 mm, 32 mm and 36 mm femoral head diameter, respectively. These results indicated a reduction range of 21.5% to 22.1% when EpUHMWPE5 replaced the existing material of UHMWPE at 50° inclination angle.

The Maximum Safe Inclination Angle graph also shows that EpUHMWPE5 exhibits the lowest total deformation with the value of 0.02546 mm, 0.02228 mm and 0.01834 mm in 28 mm, 32 mm and 36 mm femoral head diameter, respectively. The bar graph also shows that existing material of UHMWPE recorded the second highest total deformation with the value of 0.02928 mm, 0.02809 mm and 0.02306 mm for 28 mm, 32 mm and 36 mm femoral head diameter, respectively. However, only 28 mm femoral head diameter shows a reduction of 13.1% when replacing UHMWPE with EpUHMWPE5. On the other hand, the 32 mm and 36 mm femoral head diameter cases show a reduction of more than 20% in terms of total deformation.

The summarization of the result analysis is shown in Table 8.3. The data exhibits that although there are some cases in which the contact pressure recorded the highest for EpUHMWPE5, the 36 mm femoral head diameter case shows that existing material (UHMWPE) recorded the highest contact pressure. This fact proved that femoral head diameter consideration influences the contact pressure exerted at the acetabular cup region. In terms of Von-Mises stress, the EpUHMWPE5 shows the lowest value at both orientations when using 36 mm femoral head diameter. The most critical part is shown by total deformation analysis in which the EpUHMWPE5 recorded the least deformation compared to other variant materials in all cases.

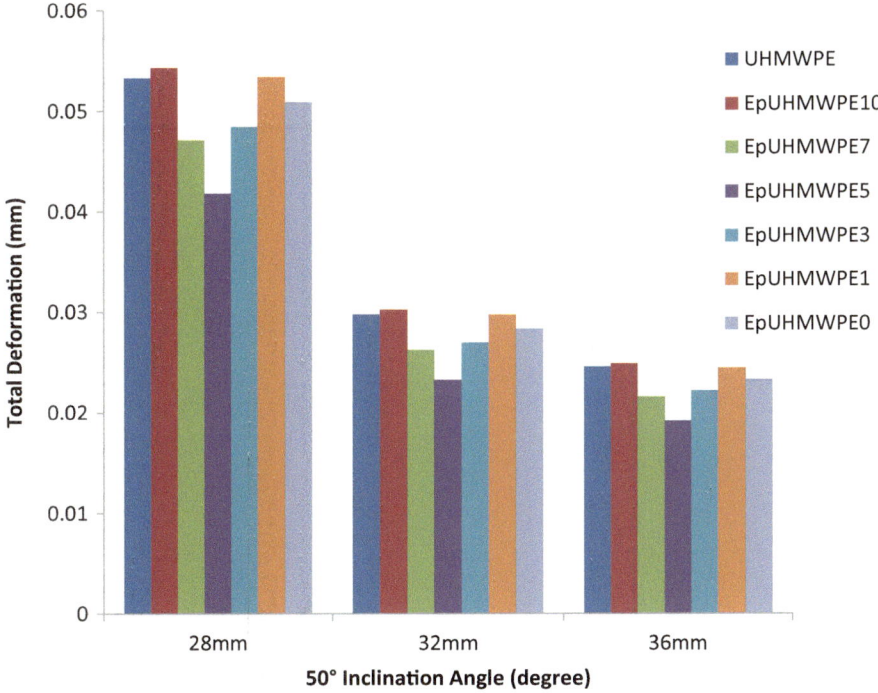

Fig. 8.6 The Total Deformation comparison of material variation in three types of acetabular components at 50° inclination angle

8.2 Summary of Chapter Eight

This chapter discussed the results of the safe zone orientation that depends on the femoral head diameter consideration. Also, the analysis using FEA exhibits that using bigger femoral head and correct positioning orientation will increase the performance of acetabular components. The new composite material study shows that EpUHMWPE5 exhibits the highest Young's Modulus compared to other variants.

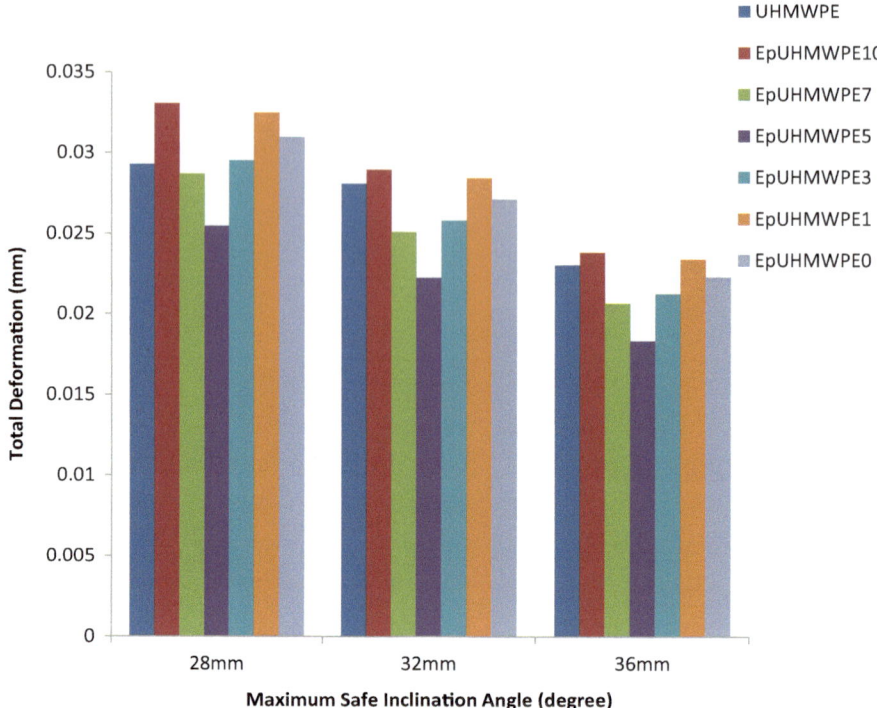

Fig. 8.7 The Total Deformation comparison of material variation in three types of acetabular components at Maximum Safe Inclination Angle (degree)

Table 8.3 The summary of FEA mechanical analysis with additional new material variants

Orientation (degree)	Head size (mm)	FEA analysis	Highest material recorded in FEA	Lowest material recorded in FEA
50°	28	Contact pressure (MPa)	EpUHMWPE5	EpUHMWPE10
	32		EpUHMWPE5	EpUHMWPE10
	36		UHMWPE	EpUHMWPE10
Maximum safe inclination angle	28	Contact pressure (MPa)	UHMWPE	EpUHMWPE10
	32		EpUHMWPE5	EpUHMWPE10
	36		UHMWPE	EpUHMWPE10
50°	28	Von-Mises(MPa)	UHMWPE	EpUHMWPE10
	32		UHMWPE	EpUHMWPE10
	36		UHMWPE	EpUHMWPE5
Maximum safe inclination angle	28	Von-Mises(MPa)	UHMWPE	EpUHMWPE10
	32		UHMWPE	EpUHMWPE10
	36		UHMWPE	EpUHMWPE5
50°	28	Total deformation (mm)	EpUHMWPE10	EpUHMWPE5
	32		EpUHMWPE10	EpUHMWPE5
	36		EpUHMWPE10	EpUHMWPE5
Maximum safe inclination angle	28	Total deformation (mm)	EpUHMWPE10	EpUHMWPE5
	32		EpUHMWPE10	EpUHMWPE5
	36		EpUHMWPE10	EpUHMWPE5

Reference

1. H. Miki, T. Kyo, Y. Kuroda, I. Nakahara, N. Sugano, Risk of edge-loading and prosthesis impingement due to posterior pelvic tilting after total hip arthroplasty. Clin. Biomech. (Bristol, Avon) **29**(6), 607–613 (2014)

Chapter 9
The Performance of Epoxy-UHMWPE at Different Inclination and Anteversion Angle

9.1 Conclusions

The safe zone orientation is very important in avoiding prosthesis dislocation and it is also proven important in MoP THR. The orientation of acetabular cup should not be generalized based on previous studies literature review. The orientation of THR must take into consideration the acetabular components size dimension, especially on the femoral head diameter. The data shows that every femoral head diameter has self-particular unique safe zone orientation. For the MoP THR, it is allowed to use the 28 mm femoral head diameter as a consideration. However, the range of safe zone orientation becomes too small. Thus, it may lead to consequences on higher margin error possibility when placing the acetabular cup based on the range safe zone orientation. The study shows that 32 mm femoral head diameter and higher will exhibit greater safe zone orientation angle for the acetabular cup.

The second conclusion was by the mechanical failure simulation studies in which bigger femoral head diameter from 28 mm will produce less contact pressure, less Von-Mises stress and improve total deformation. The results showed that transition from 28 to 32 mm femoral head diameter would reduce the contact pressure by approximately 20.5% at all angles orientation. The transition from 32 to 36 mm femoral head will reduce about 19.9% contact pressure at all orientation angles. Meanwhile, Von-Mises stress shows a reduction of 19.9% when having a transition from 28 to 32 mm femoral head diameter. However, the results showed a lesser value of 15.7% when changing from 32 to 36 mm femoral head. The total deformation improved by about 36.4% when it transits from 28 to 32 mm femoral head and 17.7% when it increases from 32 to 36 mm femoral head diameter at all orientation angles. Although the Von-Mises stress shows that all design parameters are safe, higher loading condition may induce early failure, especially on 28 mm femoral head diameter. This is due to the reason that Von-Mises stress is recorded highest at 28 mm femoral head diameter in all orientation angles. The results in FEA also indicated that anteversion angle does affect the analysis in which the 28 mm femoral head

M. F. b. A. Manap et al., *Total Hip Replacement (THR)*,
SpringerBriefs in Applied Sciences and Technology,
https://doi.org/10.1007/978-981-96-0975-8_9

diameter shows the highest reduction at Maximum Safe Inclination Angle compared to 50° inclination angle.

The third conclusion shows that EpUHMWPE could be a proposed material to replace the UHMWPE material of the acetabular cup. The range of the variants Young's Modulus shows almost identical values as compared to UHMWPE alone. Although further study needs to be made especially in vitro study and other mechanical aspect parameters, the results were promising especially for the EpUHMWPE5 variation. One added finding from here is the fact that Young's Modulus keeps increasing with added filler of UHMWPE until 5%. The value of the Young's Modulus dropped when the filler added is more than 7%. This could be caused by too much particulate of UHMWPE powder that may loosen the bonding of the new composite. However, the FEA indicated that EpUHMWPE5 could perform better than UHMWPE in terms of total deformation and Von-Mises stress. The EpUHMWPE5 featured stiffer behavior as their Young's Modulus is the highest among other variants. Although the contact pressure is slightly increased in some cases, the suggested orientation and 36 mm femoral head diameter selection could play its role in avoiding excessive contact pressure on the articulation area. This suggestion is proven by the analysis of 50° inclination angle versus the maximum angle in Sect. 4.4.2, which exhibits that EpUHMWPE5 does not necessarily produce the highest contact pressure at all types of acetabular components.

This research has successfully predicted the performance of Epoxy-UHMWPE at different inclination and anteversion angle. The results obtained from FEA mechanical analysis indicate that EpUHMWPE5 performs better compared to the existing material UHMWPE. The additional parameters of femoral head diameter and new safe zone orientation had also enhanced the performance of EpUHMWPE5 acetabular cup.

9.2   Recommendations for Future Work

On the aspect of recommendations, there are some limitations especially during the calculation of the safe zone orientation. The values 'a' and 'b' which are defined as the neck-stem angle are fixed; thus it shows that only one design parameter of THR is taken into account during the calculation. Technically, if the values 'a' and 'b' vary, all the range of safe zone orientation will be different even if they had the same head-neck ratio. In the FEA part, the pelvis model was not included in the analysis as the aim is to reduce the complexity of the project. It is important to consider the pelvis model as a part of the FEA analysis as it will increase the accuracy of the analysis that was conducted. There are two main mechanical data important for FEA analyses, which are the Young's Modulus and Poisson's ratio. In this study, we make the assumption on the value of the Poisson's ratio of EpUHMWPE variant due to the small magnitude and less than 10% filler added. The Poisson's ratio experimental process will allow the exact value of every variant and also allow adding more variation percentages into this project.

Index